Urban China Reframed

Given China's rapid economic growth and massive urbanization, no one in the world can ignore what is happening in urban China. This book is a critical review of existing urban China research, which is found wanting due to the decontextualized use of theories and concepts developed in the West.

Urban China Reframed: A Critical Appreciation consists of epistemological, theoretical and methodological contributions to remedy these limitations by focusing on a number of relevant topics. First, models are widely employed in any study, and China nowadays has invoked models like city system, zones and global city in socio-economic development. How to interpret them in terms of knowledge production in a strong party-state? Second, given the global prevalence of neoliberalism, it is an important debate whether neoliberalism is applicable to China. Third, what is urban ideology in China? How to contextualize it? Are debates about the differentiation between the city and urbanization relevant to China? Fourth, massive rural-urban migration in China has taken place within its mega rural-urban dual system, an institution that has persisted since the 1950s. How does it manifest nowadays? Fifth, has the town-country divide in China, like in the West, disappeared? If not, how can one interpret China's town-country relations, within the politics and administration of the Chinese state? Sixth, how to decipher the territorial development in the Pearl River Delta, the "world's factory," under the auspices of the state? The collection of essays in this volume contributes to the theoretical understanding of urban China.

The chapters in this book were originally published in the *Eurasian Geography and Economics.*

Wing-Shing Tang was formerly Professor at the Department of Geography and Research Fellow at Advanced Institute for Contemporary China Studies at Hong Kong Baptist University, China. His research interrogates urban theories in China, with special emphasis on town-country relations.

Kam Wing Chan is Professor of Geography at the University of Washington, Seattle, USA. He specializes on China's urbanization, migration, and the household registration (*hukou*) system.

Urban China Reframed

A Critical Appreciation

Edited by
Wing-Shing Tang and Kam Wing Chan

Routledge
Taylor & Francis Group
LONDON AND NEW YORK

First published 2021
by Routledge
2 Park Square, Milton Park, Abingdon, Oxon, OX14 4RN

and by Routledge
605 Third Avenue, New York, NY 10158

Routledge is an imprint of the Taylor & Francis Group, an informa business

British Library Cataloguing-in-Publication Data
A catalogue record for this book is available from the British Library

ISBN13: 978-0-367-70645-6 (hbk)
ISBN13: 978-0-367-70647-0 (pbk)
ISBN13: 978-1-003-14736-7 (ebk)

Typeset in Myriad Pro
by codeMantra

Publisher's Note
The publisher accepts responsibility for any inconsistencies that may have arisen during the conversion of this book from journal articles to book chapters, namely the inclusion of journal terminology.

Disclaimer
Every effort has been made to contact copyright holders for their permission to reprint material in this book. The publishers would be grateful to hear from any copyright holder who is not here acknowledged and will undertake to rectify any errors or omissions in future editions of this book.

Contents

Citation Information

The following chapters were originally published in the *Eurasian Geography and Economics*, volume 60, issue 4 (2019) and volume 58, issue 3 (2017). When citing this material, please use the original page numbering for each article, as follows:

Chapter 1
Exemplary Cities in China: The Capitalist Aesthetic and the Loss of Space
Carolyn Cartier
Eurasian Geography and Economics, volume 60, issue 4 (2019), pp. 376–399

Chapter 2
Uncorking the neoliberal bottle: neoliberal critique and urban change in China
Will Buckingham
Eurasian Geography and Economics, volume 58, issue 3 (2017), pp. 297–315

Chapter 3
China's urban ideology: new towns, creation cities, and contested landscapes of memory
Tim Oakes
Eurasian Geography and Economics, volume 60, issue 4 (2019), pp. 400–421

Chapter 4
Two systems in one country: the origin, functions, and mechanisms of the rural-urban dual system in China
Kam Wing Chan and Yanning Wei
Eurasian Geography and Economics, volume 60, issue 4 (2019) pp. 422–454

Chapter 5
Town-country relations in China: back to basics
Wing-Shing Tang
Eurasian Geography and Economics, volume 60, issue 4 (2019) pp. 455–485

Chapter 6
Territorially-nested urbanization in China – the case of Dongguan
Kit Ping Wong
Eurasian Geography and Economics, volume 60, issue 4 (2019) pp. 486–509

Contributors

Will Buckingham Formerly, Department of Geography, University of Washington, Seattle, USA.

Carolyn Cartier School of International Studies, University of Technology Sydney, Australia.

Kam Wing Chan Department of Geography, University of Washington, Seattle, USA.

Tim Oakes Department of Geography, University of Colorado Boulder, USA.

Wing-Shing Tang Formerly, Department of Geography and Advanced Institute for Contemporary China Studies, Hong Kong Baptist University, China.

Yanning Wei School of Environmental Sustainability, Loyola University Chicago, USA.

Kit Ping Wong Future Cities Laboratory, Singapore-ETH Centre, Singapore.

Reframing urban China research: a critical introduction

Wing-Shing Tang

Introduction

Almost half a century has passed since a few of the earliest monographs on urban development in post-reform China were first published in English (e.g., Kirkby, 1985; Ma and Hanten, 1981). Since then, urban China has attracted plentiful attention in the English language literature, even surpassing its bigger brother, Soviet Union/Russia. For example, two post-socialist urban geographers have made an estimation on the number of papers published in *Eurasian Geography and Economics*, the successor of *Post-Soviet Geography and Economics* (and the latter itself is the successor of *Soviet Geography*), and found that for the decade since 2005, there were a total of 26 articles roughly classified as urban geography (and there are many more on other themes), with 12 and 5 on Chinese and Russian cities, respectively (Ferencuhová and Gentile, 2016, 485). It is the objective of this introductory chapter to critically comment on this development to lay the foundation for the rest of the book. Having evaluated this burgeoning literature by two related concepts of random conceptual indigenization and random conceptual appropriation (Tang, 2014a), this chapter summarizes the contents of various chapters and highlights their possible contribution to our understanding of urban China.

'The Urban Revolution' in China

In a paper presented in a conference organized by Neil Smith almost 15 years ago,[1] I postulated that the then experience of urban development in China can also be conceptualized as 'the urban revolution'.[2] By the mid-2000s, urban development in China had already exhibited at least four characteristics. Urban growth has continued to be extensive in nature; the economic functions of urban areas have been enhanced; urban areas have more concerns of their own; and urban areas are increasingly socially stratified and unequal. To confront new challenges, local governments have been empowered to control on behalf of the vanguard state.[3] Besides, the state has started to conceive of individual agents as enthusiastic producers and consumers as well as patriots. Accordingly, there is the proliferation of new technologies of government, especially 'territorialized technologies', on top of those persisting, moral technologies.

Of course, this understanding of 'the urban revolution' at that time is still different from the one as first coined by Lefebvre in his famous treatise, *The Urban Revolution*, in which Smith wrote an introduction in the English translation. For Lefebvre, the revolution refers to the development from city to urban society and so the concern of urbanization, and the shift in the contradiction from between town and country to that between center and periphery. China had, however, reached 'the urban revolution' in the early 2000s in a distinguishable path. According to Tang (2019), in the dynastic past, *cheng* (the Chinese concept of city) was never an independent unit, given the mutually embedded *cheng*-cum-*xiang* (equivalent to town-country) relations. While *cheng* had occupied a more prominent position in the Republican period, it had still exhibited characteristics very different from the West. Guangzhou, for example, was established as a *shi*, for the intention of promoting provincial independence. Then, *shi* had never obtained an independent logic during the Maoist regime, as it was only a center in the nationally organized production network, a node within the networks of government and a site of socialist representation. It was on the basis of this *longue durée* that we can start to decipher the aforementioned 'urban revolution' and understand accordingly why and how the Chinese is different from that theorized by Lefebvre.

The 'evolution' in urban China research

Against this backdrop, it is timely to review how urban China research has deciphered 'the urban revolution'. Given that growth rates of China even by the mid-2000s had been, according to the economist, Peter Nolan (Nolan, 2004, 9), at the top of the world's growth performance during that period, its massive urban growth has, obviously, attracted abundant intellectual attention. The literature has then grown by leaps and bounds, and two interrelated trends can be identified.

First, since the country opened up itself to the world again at the end of the 1970s, the general trend was the distancing away from the Soviet influence on many fronts, although the traditional Chinese (informed by the legalist philosophy) emphasis on practical rationality and the Maoist emphasis on practicing class struggle based on friends and foes had somehow twisted the Soviet impacts in the first place.[4] The way that *shi* has been treated intellectually is no exception to this general rule. The knowledge bases of territorial production systems and locational principles, derivable from the Soviet central planning system, emphasized knowledge about prospective development regions. It was the allocation of all sorts of resources, including the migration of people, across space that occupied the top of the priority. There was a comparatively minor role in the knowledge about cities with independent logics of development. Everything had changed since the opening-up at the end of the 1970s. The state has practiced an updated rationality that encourages enthusiastic agents, including peasants, foreign capitals, some regions and consumers, to take economic developments to new heights. The multiplicity of spatio-temporalities of these agents has demanded new knowledges of spatialization and spatial technologies. Besides, the earthquake in Tangshan had urged people to query whether there was a limit to the city. The rapid pace of urbanization alerted the national economic planners to problems like the relationship between the level of urbanization and the level of economic development; rural-urban migration; the urban-size hierarchy; and the functional structure, catchment area and spatial structure of *shi*. As a result, *shi* is then expected to take into consideration of relations with the countryside for annexation purposes; foster entrepreneurial calculability in the increasingly interconnected world; concern itself with the dynamics of its economy, its population and its environment; and regulate the conditions underlying the burgeoning real estate market. The complexity of the urban question led to the birth of urban studies in China (e.g., Tang, 2006; Tang, forthcoming).

Second, at the early stage of the opening-up, it was the ethnic Chinese, who, in interaction with the outside world, have transmitted these messages about the changes in development. As time proceeded, when an increasing number of non-Chinese started to cross the language barrier as well as learn more about the development via fieldwork or the literature, they have also played a role in interpreting these changes. A classic example of the latter is the study on Chinese urbanization written by Richard Kirkby (1985), who got acquainted with the development while teaching English as a foreign language in the mainland and drawing on local students' translation of Chinese studies on various aspects of urbanization into English for his interrogation. The other example is Guldin (1992), who has helped re-invigorate the field of urban anthropology in China.

Regarding the mainland Chinese contribution to urban studies, it is imperative to mention the role of the urban geography discipline. Given its basic contribution to the regional foundation of urban development, urban geography has been the dominant sub-discipline within urban studies at its earliest stage of development. Urban geographers have started to investigate a whole array of issues such as land assessment, 'commodity' (i.e., non-state) housing, urban transport, urban redevelopment, suburbanization, the impact of foreign direct investment, urban governance, world city formation and spatial inequalities. Starting the early 2000s, with the growth of the 'middle-class' as well as poverty in the city, attempts have begun to explore the cultural turn of urban issues as well as urban enclaves at the fringe (Tang, forthcoming). By the second decade of the 21st century, given the increasing number of Chinese scholarly writings in English, there has even emerged a number of review papers, trying to recount the progress and propose newer pointers for future developments (e.g., He and Lin, 2015; He and Qian, 2017; Huang, 2020; Wu, 2020). It is fair to generalize from this huge literature (one survey claims that there were altogether around 2,500 articles published between 2003 and 2013 (Jayne, 2018, 6)) that given the nature of reform being marketization, one form or other, within the constraint of the party-state, the common preoccupation of this literature is to document the urban manifestations, as well as their corresponding technologies (such as urban planning and administration), of the conflicting state-market and state-society relations. Lately, due to the further exposition to the Western mainstream literature, there have been calls to orient the urban China literature towards more politicized, more social, and more environmentally sensitive deliberations. Some have already attempted to destabilize the mainstream literature by confronting popular topics, such as variegated neoliberalism (e.g., Zhou, Lin and Zhang, 2019) and the right to the city (Qian and He, 2012). Still others even conjecture to contribute to the development of urban theories that emerge from the Global South (He and Qian, 2017, 840–841; Ren, 2018). Wu (2020, 469) advocates practicing Jennifer Robinson's comparative urbanism approach: "stick to original observations from the field and to add the narratives to expand our imagination of the urban". Zhou, Lin and Zhang (2019, 40) argue that urban China is even a site of original theory production. The thinking in and about China has

> to reach across time and across space seeking critical engagement with a general audience beyond China. ... there is no reason why theorisation of Chinese practices and urban experiences cannot find a way to 'theorise back' to the hegemonic power of the West.

There is the other side of the story as that is contributed by the Western mainstream urban scholars. Well-informed of various theories and schools of mainstream urban studies, these scholars tend to comment on the progress of the urban China literature accordingly. One hot topic of debate in the West is whether China's urban development is exceptional or generalizable to Western urban theory. Having evaluated on the basis of five aspects (scale, speed and extent; the state; rural land acquisition; redevelopment and gentrification; and the market), Hamnett (2020, 697) came to the conclusion that China has exhibited "a very different form of urbanisation" and it is, alas, the most popular remark, "urbanisation with Chinese characteristics". Jonas (2020) takes up the debate more specifically by employing Allen Scott's city-regionalism as the yardstick to evaluate the recent proliferation of studies on city-regionalism in China. He concludes that "China's urban development processes are neither exclusively unique nor fundamentally resistant to theorisation. They are in this respect essentially comparable and thus highly amenable to theoretically informed generalisation" (Jonas 2020, 707). There are others who are anxious to provide prescription for the urban China literature. One example would be that furnished by Jayne (2018, 1), who asks the question: "why is it that scholarship focused on urban China is not making more of an impact as cutting-edge, innovative and adventurous urban thinking?" The response appears in Hamnett (2020, 697): due to the proliferation of "very technical and detailed quantitative analysis" in the field. This is understandable as most geographers in China have backgrounds mostly in science and engineering, and many geographical ideas have been introduced with the objective of ensuring state governing (see below; Tang, forthcoming). For Jayne (2018, 4), the issue is that "there has been a relative lack of engagement with critical theoretical perspectives – including Marxist, feminist, post-structural, post-colonial, and 'more-than-representational' thinking". The West should offer help to develop critical scholarship on urban China, even though what he has in mind is the comparative urban research approach, rather than the Chicago or Los-Angeles types. Finally, there are attempts to advise the employment of individual concepts to understand urban China. Of course, the most renowned of all would be Harveys' (2005) coinage of 'neoliberalism with Chinese characteristics'. Given his prominent position, application of this concept has been applied profusely to urban China. Even when the vanguard state summoned to develop Marxism again in mainland China in a Philosophy and Social Sciences Work Conference in celebration of the bicentennial of Marx's birth, in May 2015, Harvey was called upon to deliver talks and, subsequently, engage with local academics in June 2016 (Cai et al., 2016a, b; Tang, forthcoming). Other concepts include glurbanization and the interscalar advocated by Matusitz (2010, 21), who argues that it is "essential to analyze urban transformation in East Asia from an interscalar perspective" and "as a world city, Shanghai has all the features of a US city". Similar calls have been made to concepts like enclave urbanism (Douglass, Wissink and van Kampen, 2012; Wissink, 2019) and gentrification (Shin, 2016). Nowadays, there are many more debates of this kind in English language journals, but the few aforementioned examples have addressed the nature and contents of the debates.

The poverty of the literature: random conceptual indigenization and appropriation

In their reviews, people like Hamnett (2020) and Jayne (2018) do recognize that the Western mainstream urban studies has over the years developed from, just to name a few, the Chicago school, the Marxist, the post-modern, the post-structuralist, the post-colonialist to the assemblage. These trends have, however, been documented, or even evaluated, like fashions that come and go. While the sub-altern, post-colonial and de-colonial camps (SAPC) may be more emphatic in emphasizing the power relations involved in the production of these strands and their impacts on the under-privileged, others have basically kept silent. In pushing forward their arguments, SAPC have, somehow, dichotomized the world into dualities of North and South, or History 1 and History 2 (Lazarus and Varma, 2008). In doing so, they have missed the mutual embeddedness between these two opposing forces, as informed by the *tongbian* philosophy (Tang, 2019, 2020). There is no exception to the interrogation of knowledge production. More specifically, it is difficult to merely put all the blame on the developed world while eschewing the intervening role of the developing world. Even within the developing world, it is equally true not to forget about the role of government as well as the academics in local knowledge production. There is a need to articulate knowledge production via related histories of local, national and global development. Knowledge production nowadays is in fact articulated with practices in the past as well as those in another place at a distance. These complexities cannot be revealed by merely examining publications and citation patterns, as Kong and Qian (2019) have done to reveal Anglo-American hegemony in urban studies, with an additional emphasis on urban China research. The key to understanding these complexities lies elsewhere.

Earlier, I (Tang, 2014a) have coined two interrelated concepts of random conceptual indigenization and random conceptual appropriation to decipher the nature of research found in the urban China literature. Couched in Smith and Katz's (1993) discussion of spatial metaphor, the two concepts can be schematically represented in Figure 1.

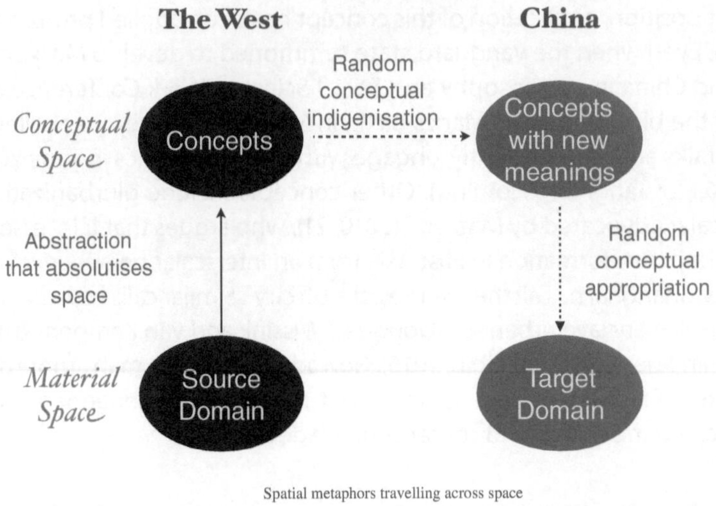

Figure 1. Schematic Representation of Random Conceptual Indigenization and Appropriation.

The former concept refers to the double randomness in producing knowledge about urban China space. Researchers have adopted the spatial container approach of understanding that ignores the processes and context of knowledge production. It is usually assumed that a Western concept, which is free-floating, can be applied to China indiscriminately, "without reference to the material practices and actual spatial experiences within that space and relative to and in relationship with other spaces in the West" (Tang, 2014a, 48). Absolute space has been privileged at the expense of relative and relational space. In the process of indigenization, one usually finds an additional twist performed by many urban China researchers: adding a signifier (the commonest of all is 'state-led'[5]) to the Western concepts to emphasize that the concept contains Chineseness, not Westernness, underscoring their 'area studies' contribution. Examples of merely adding signifiers to Western concepts are prolific as shown in Table 1. Although the table has only summarized a survey undertaken up to the mid-2010s, the message of abundant random indigenization in the literature is still loud and clear. To sum up, underneath random indigenization is the power relation of knowledge production between the hegemony of the West and the non-West.

In contrast, random conceptual appropriation refers to a different power relation: between researchers and their own governing bodies. Very often, concepts introduced by researchers for the benefit of governing may turn into rhetoric to rationalize policies. "[R]andomness is defined by the scrutiny of the concept according to the particular requirement of governing, irrespective of knowledge production in the first place" (Tang, 2014a, 50). Solinger and Chan (2002, 204) made a similar remark about China urban studies almost two decade ago: "[T]he China urban studies field has been little affected by scholarly fashion, but much influenced by Chinese politics and state policy". Such kind of random appropriation is best exemplified by the Action Area Plan for the Bay Area of the

Table 1. A Glossary of Random Conceptual Indigenization in the Urban China Literature

Western Concepts	Randomly Indigenized Concepts Employed in Urban China Research
State-society relationship	*Strong* state-society relationship (Lin, 2007)
Growth coalition	*The Chinese* growth coalition (Wu, Xu and Yeh, 2007, 70)
Self-help housing	Self-help housing *in the dual housing market* (Zhang, Zhao and Tian, 2003); *Chinese* informal settlements (Wu, Zhang and Webster, 2013)
Gentrification	*State-sponsored* gentrification (He, 2007); *state-led* planning approach to 'villages-in-the-city' (Chung, 2009)
Property rights	*Ambiguous* rights (Zhu, 2002); *reduction* in transaction costs to land dispossession (Zhao and Webster, 2011)
Gated communities	*Foreign* gated communities (Wu and Webber, 2004)
Neoliberalism	*Localized* neoliberalism (He and Wu, 2009); *municipal-government-led* urbanization of capital (Lin and Yi, 2011); *variegated, polymorphic* capitalism (Peck and Zhang, 2013)
Neighborhood effect	Neighborhood effect *of the market and state institutions* (Wu, He and Webster, 2010)
Social production of space	*A particular* social production and social construction of space (Zhang, 2011)

Pearl River Delta Estuary in Hong Kong in 2011 (Figure 2). The technical team (including Hong Kong, Macau and Guangdong provincial governments) drew on Athens Charter and the UN Inhabitation Agenda (both referring to cities), and practical experiences of world-renowned livable regions, such as San Francisco Bay Area and Tokyo Bay Area, to formulate the concept of livable region. But this concept was mutated by the Construction Ministry of China's livable cities scientific evaluation index system during the compilation of the basic ingredients of Bay Area as a livable region. Knowledge was, first, appropriated from that about city in the West, then enlarged it to that about region, and, finally, catered for the governing of a territorial region in China. Underneath this technical exercise was the state's interest in pursuing the 'Bay Area strategy' as a means to integrating Hong Kong into the mainland (Tang, 2012). This randomness in appropriation is of utmost importance for our understanding of the power relation in knowledge production, as it allows us to avoid falling into the trap of blaming the West solely on all the conspicuous problems, as SAPC as well as a Marxist analysis of hegemony would have us believe.

These random attempts in the literature have produced many problems in the understanding of urban China. There are epistemological problems of relating sources of knowledge solely to the West. To overcome this, we must downplay the absolute concept of space as well as uphold the relative and relational concepts by paying more serious attention to processes of development in China. Theoretically, many popular concepts in the urban China research literature must be reviewed thoroughly. In particular, the Chinese vanguard state must be theorized in terms of its rationality of intervention, power and politics in particular, rather than merely concentrating on the apparent changes in institutions and agency, a task only sufficient for the purpose of identifying and depicting the signifiers. Nor is it enlightening at all to rely on irrelevant concepts like neoliberalism or governing without

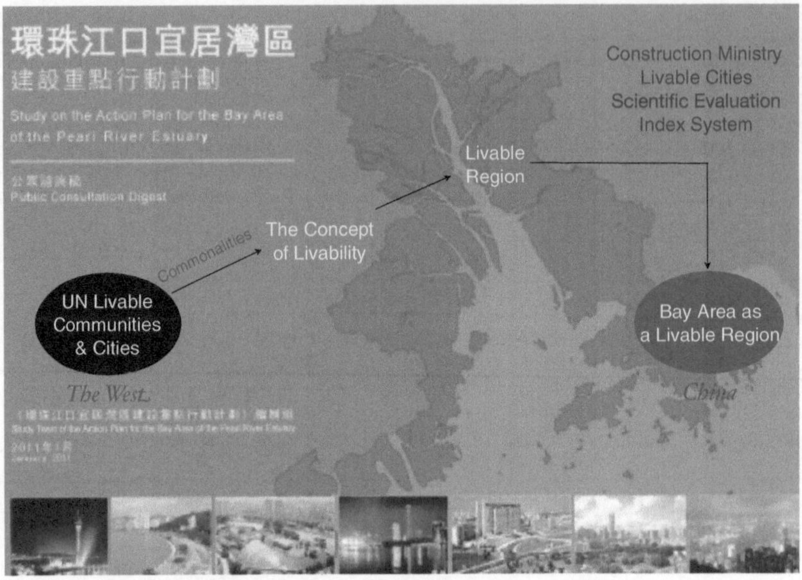

Figure 2. An Example of Random Conceptual Appropriation: The Bay Area.

government. Methodologically, one must concentrate not simply on the most recent developments at the expense of the past. Dividing post-1949 China into two mutually exclusive pre- and post-reform periods is absolutely a hindrance to penetrating the reality. An historical approach to trace the developments over time is desperately needed, in addition to a spatial one deciphering the intricate interaction between China and the world. As a result of these potential improvements, politically, urban research will equip us with the necessary analytical tools to stop de-politicizing many urban issues, thereby eschewing crushing the everyday lives of people. Cities will then become more just, fair, less alienated, less censored and supportive of everyone who lives in them.

Coining these two concepts to expose the deficiency of urban China research and then suggesting possible improvements is one thing, how to actually do it is another. Once the 'pandora's box' is opened, many more issues do arise. What are really the appropriate concepts? How should we proceed to derive them? Armchair conceptualization or painstaking field research? How essential is the past to our understanding of the urban present? Similarly, what is the role of the Chinese setting in comparison with the outside world? Which is the outside world, the West, the Soviet Union or what? The rest of this chapter summarizes the preliminary attempts by concerned scholars to reframe the literature.

The contents of this book

In this volume, Carolyn Cartier (2019) substantiates the concept of random conceptual appropriation through an examination of paradigms commonly adopted in urban China geography based on the history of the capitalist city and liberal political economy. She enlists Thomas Kuhn's critical interpretation of exemplar paradigms to demonstrate how applications of concepts such as city system, zones and global city to China often depend on bracketing Chinese political conditions to accommodate the theory. The result produces a 'capitalist aesthetic' of urban appearances and a 'loss of space' of those contexts that scholars ignore because they do not 'fit' the Western model. Cartier also troubles this argument with Børge Bakken's provocative analysis of science and scientific models in China's socio-economic development, which informs how the party-state promotes normative scientific models in relation to attainment of university rankings and other world standards.

Will Buckingham (2017) interrogates the largely uncritical adoption of neoliberalism as an explanatory device for China's development. He finds that neoliberal critique fails to adequately explain China's spatial development because it assigns causality for social and economic inequality to globalized processes of capital accumulation while resigning Maoist institutions to quirks of 'actually existing neoliberalism'. The result is a framework that obfuscates both the mechanisms of China's urbanization and the processes through which neoliberalism is adapted, adopted and resisted in the formation of urban policy and planning. He, too, draws on the concepts of random indigenization and appropriation to argue that urban studies must look beyond the superficial ways in which new spaces 'appeal to capital'. Instead, urban scholars should produce research that is grounded in the real histories of place and the lived realities of daily life.

Tim Oakes (2019) put forward two conceptual distinctions regarding urbanization for urban China research. First, it relates to the differentiation between the material basis of cities for the country's economic development in the global economy, and the concern for state-building and party-power. For him, urban ideology should be understood in the latter sense. Second, while debating with Lefebvre in that one should treat the Chinese case not so much the empirical object as an object for its own theory building, Oakes has approved the Lefebvrian distinction between 'the city' and 'urbanization'. Cities, their historical process in the making as well as their material artefacts, including urban planning, are constantly deployed by the Chinese state to control urbanization.

Kam Wing Chan and Yanning Wei (2019) elaborate the continuing growth of the rural-urban dual system in China from the Maoist period to the present. The socialist strategy of economic development had in the past established the *hukou* system, restricting surplus labor in the countryside from migrating to the city and, in corollary, the consolidation of the dual system. Against the paradigms of market transition, neoliberalism and migrant assimilation is, they argue, the historical continuity and pliability of the dual system in the last three to four decades. Nowadays, the dual system has related more to land seizures: it has rendered migrant labor cheap for the world factory and allowed the state to expropriate rural land.

Wing-Shing Tang (2019) queries the understandings of town-country relations in China. Much of the existing research proposes a dualistic understanding, which is largely drawn from the West. To redress this imbalance, he advances a nuanced understanding based on a methodology of spatial story, underscoring spatio-temporal relations; an epistemology, informed by *tongbian* philosophy, of understanding relations not so much dualistic as mutual embedded between polarities within a focus-field relationship; and a theorization of the town-country relations as *cheng*-cum-*xiang* relations in dynastic empire-building and its spatialization. He also elaborates these relations by illustrating their historical development for the city of Guangzhou.

Kit Ping Wong (2019) has challenged the commonplace understanding of Chinese urbanization in the literature: a dualist thinking of either 'bottom-up', 'village-led' or 'state-led' processes. She argues, however, that it is imperative to put forward a political modality of territorialization through the interrogation of the intertwined relationship among state power, urbanization and territorialization. In particular, she coins the concept of 'territorially-nested urbanization'. Through different modes of territorialization, the party-state controls the hierarchization and differentiation of urban processes, orchestrating territorial dynamics between concentration and extension of urbanization. Under the 'territorial nesting' of state space, different modes of state power can operate simultaneously from various ranks of urban territories. These complicated arguments are elaborated as well as contextualized by three related levels of territorialization: Guangdong province, the city of Dongguan and the field analysis of individuals at the concrete ground.

Contributions of this book

This volume makes a few contributions in epistemology, methodology and theory. In epistemology, all chapters have emphasized the obvious, yet often ignored, fact that the process and context of knowledge production basically reside in China. For example, Cartier traces scientific exemplarity through the development of geographical societies since the beginning of the last century and ways that geographical research has accordingly privileged model-building. Chan and Wei have related the roots of the present dual society to the development strategy of Maoist China back in the 1950s and Soviet Union in the 1930s. Tang has pushed their argument further by linking up what happened during Maoist China with the dynastic past (and then the Republican era). Hence the sub-title of his chapter: "Back to Basics". Wong and Oakes enlighten us that Chinese urbanization has a historically specific logic of political modality of territorialization and of party-state building, respectively. Buckingham insists that understanding of China must be grounded in the real histories of place and the lived realities of daily life.

Having made this remark, the chapters also recognize the necessity of underscoring the interaction of China with the rest of the world. Thus, methodologically, they trace the phenomena under investigation not only diachronically to China's past but also synchronically with the world. In particular, the senior author of Chan and Wei is the ardent advocate, if not the first one, who promotes the study of Chinese phenomena by tracing them back to the Soviet Union. His contribution can be found from his earlier classic, published in Chan (1994), *Cities with Visible Walls: Reinterpretation of Urbanization in Post-1949 China*. Obviously, China learnt socialism and socialist practices from the Soviet Union. Adopting *hukou* is one example, and the economic development strategy is another. Chan's contribution to urban China research is that instead of looking to the West for analytical insights, one should turn to the Soviet practices to avoid random conceptual indigenization and appropriation. To respond to this call, Tang (2019) has even proposed a formal methodology, the spatial story approach, to decipher the interaction between China's past and its interaction with the rest of the world, including the Soviet Union.

Wong has contributed to this 'enlightenment' debate from another perspective: the emphasis on fieldwork. It is common for urban China researchers to go to the field for a short period of time and pick up some indicators of 'state-led' or similar signifiers. This could easily be achieved given the 'old-boy' network of many of these researchers, who happen to know one or more friends or students in the government offices in the case study site. The merits of this kind of fieldwork are, dialectically, its demerits too. Since they can get hold of data and information easily, these researchers seldom ask deeper questions about the phenomena under study, and gloss over the underlying causes. Then, this kind of end-oriented, appearance-biased, short-term geographic fieldwork, while informative in some sense, cannot be a substitute for a deep, anthropological fieldwork, which requires researchers to live there, engage with people, mostly the ordinary citizens other than those in prominent power positions, for insights underneath the phenomena portrayed by the former field approach. In her own words (Wong, 2019, 488):

> This fieldwork constitutes an important part of the study by acknowledging the urban as a mediation between diverse levels of force, processes and relations. It will become clear that state power does not operate in a clear, coherent or homogeneous way, but is rather opaque, multiple, and contradictory.

Having carried out this kind of engaging fieldwork, Wong has been able to interrogate the material manifestation of the politics of urbanization in the urban China literature, which is difficult to be portrayed by dualistic arguments.

Theoretically, the chapters in this special issue have explored many new concepts with great potentials. In the mainstream urban research, there is the concept of 'loss in translation' to denote the loss of original meaning in the process of translation. Cartier, drawing on Kuhn and Bakken, and many others, has instead coined the more geographic concept of 'loss of space' to characterize the urban China literature that is fond of formulating models that limit the full range of understanding of the urban processes in China. In her words, "There is an annihilation of geography by space" (Cartier, 2019).

The commonplace practice of random conceptual indigenization and appropriation is to join the bandwagon of the Western literature by indigenizing or appropriating these concepts to urban China. In these days, almost no one in the academic world on space is more popular than Lefebvre, especially his treatise, *The Production of Space*, which has been applied worldwide. No wonders that there are many research papers uncritically drawing on Lefebvre to understand urban China (see Tang, 2019). Two of the chapters in this volume, Oakes and Tang in particular, have, however, taken issue with Lefebvre. While finding Lefebvre useful to understanding 'the city' as an ideological project of spatial control, Oakes also notes that the inevitability of urbanization underlying Lefebvre's theory of urban revolution is based on assumption of rural-urban antagonism that describes the emergence of capitalism in Europe but that does not fit well with China's history at all. Tang (2014a, b, 2016, 2017, 2020, Forthcoming) has been critical of applying his trialectics, concept of representation, thesis of complete urbanization and the transition of town-country to center-periphery contradiction, to Hong Kong and China. In the chapter in this book, he has focused more specifically on the conceptual understanding of town and country in China. Specifically, he underscores the mutual embeddedness between the two. As this characteristic must be attributed to the systemic characteristic of dynastic empire-building, which is absent in the West, Tang echoes Faure and Liu's (2002, 2) observation that "what the educated Westerner might mean by 'city' or 'village' does not correspond to what the educated Chinese person means by *cheng* or *xiang*" by coining the more relevant concept of *cheng*-cum-*xiang* to capture the subtleties of the mutual embeddedness between *cheng* and *xiang* in China. Although Wong did not explicitly mention the contribution of Lefebvre in her chapter, one can still detect his influence. Her concept of extended urbanization has his imprints, but, more powerfully, it has been expanded in correspondence with the China's party-state's territorialization, again yet intervened or circumvented by numerous attempts of towns and villagers on the ground.

The other concept that has been tackled more seriously in this special issue is that of the state. As aforementioned, it is common for the urban China literature to simplify theoretical interrogation by adding the signifier of 'state-led' to the concept of state as if the Chinese was then different from the West. All chapters in this volume have rejected these random attempts by dwelling in greater details on the nature, rationality, ideology, structure, mechanisms, strategies and actions of this ontological body called 'the Chinese state'. Although Tang focuses on the dynastic empire, he has repeatedly implied that its nature, words and deeds can also be found nowadays. In particular, he argues that Lebowitz's (2012) concept of vanguard state under real socialism is more instrumental as well as perceptive to decipher the party-state in China since 1949. Unlike the 'state-led' argument that emphasizes on state ownership *per se*, the vanguard state stresses the essential operation of democratic centralism: a bureaucratic hierarchy in which commands are handed down from the top and carried out by the subordinates. To Lebowitz (2012, 82, emphasis original), state ownership must be understood in vanguard form: " ... a party of the vanguard-type, state ownership in its vanguard-form, state coordination in its vanguard-form, central planning of a vanguard-type, social and civic organizations of a vanguard-type, and, of course, an underlying population that *accepts* all this". This concept of the state, which diverges from the Westphalian state, has great implications for the concept of territory and the related spatial technologies too. The complexities of the leveled hierarchical administrative system and territorially organized ones can be elaborated by, as suggested by Tang (2019), the dynamic mutual interactions between focus and field. Cartier contributes to this call by investigating the different ways in which the party-state has promoted and debated models, including models of regional development. Of course, she herself has contributed to the debates in her other contributions on urbanization (e.g., Cartier, 2013, 2015, 2016, 2018). Buckingham joins this debate by challenging the relevancy of the superficial concept of neoliberalism to China. Oakes focuses on the production of ideology within the context of state-building and party-state power. Chan and Wei enlighten us on the formulation of development strategy and its linkage with *hukou* system. Wong insists that we should focus on territorialization, a form of practicing territory at variant with the West, which should include many other agents on the ground.

Against the backdrop of these contributions, the chapters have subtly highlighted the values of sound urban China research. Once we have (re)politicized the state, we can repoliticize many urban issues and, at the end, improved the everyday lives of people. This reminds us of one of Marx's famous remarks: "The philosophers have only interpreted the world, in various ways. The point, however, is to change it".

Some concluding remarks

This book has preliminary challenged "many orthodoxies" that have plagued the field and then reframed the literature by opening possible realms of critical China urban studies grounded in history and reality, yet built on the literature developed in the geographical East, the ideological East and the West. May random conceptual indigenization and appropriation in the field perish. Long Live Urban China Research!

Acknowledgments

I would like to thank all other authors in the book for their comments on this chapter. Remaining errors are, however, mine.

Notes

1. This paper, entitled "Chinese Urbanism Today: *La Révolution Urbaine?*", was presented at the Conference "The New Chinese Urbanism", Center for Place, Culture and Politics, Graduate Center, City University of New York, 23–24 May, 2005. A version of it was later translated into Japanese and published as Tang (2007).
2. In the conference paper, the concern is whether China has experienced a phenomenon like 'the urban revolution' as raised by Lefebvre, while the answer is negative. In retrospect, at the end of 2020, as well as given my own critique of the urban China literature (Tang, 2014a, 2019), I should have had coined a more contextual concept than 'the urban revolution' in 2005. But this is part of a learning curve.
3. The concept of vanguard state is drawn from Lebowitz's (2012) characterization of the Soviet Union.
4. To say the least, the central planning system practiced in China in the 1950s was "highly aggregated and ill-defined, covering a much smaller number of products than in the Soviet Union and rendering the system more open to negotiation" (Tang, 1997, 37). "[T]he Chinese urban land system had evolved out of the historical necessity of placating the peasants in the rural areas and the industrialists in the urban areas in the early 1950" (Tang, 1994, 394).
5. This practice of adding the signifier of 'state-led' is not exceptional to China, but can be commonly found elsewhere. An example would be Sonn and Shin's (2020, 869) study of developmental state-led gentrification in Seoul, South Korea.

References

Buckingham, Will. 2017. "Uncorking the Neoliberal Bottle: Neoliberal Critique and Urban Change in China." *Eurasian Geography and Economics* 58: 297–315.

Cai, Yunlong, et al. 2016a. "Makesi Zhuyi Dilixue jiqi Zhongguohua: 'Kuaguo, Kuojing, Kuadai' Zhishi Xingdong [Knowledge as action: Marxist geography and its development in China]." *Dili Yanjiu [Geographical Research]* 35: 1205–1229.

Cai, Yunlong, et al. 2016b. "Makesi Zhuyi Dilixue jiqi Zhongguohua: Guihua yu Shijian Fansi [Marxist geography and its development in China: Reflections on planning and practice]." *Dili Yanjiu [Geographical Research]* 35: 1399–1419.

Cartier, Carolyn. 2013. "What's Territorial about China? From Geopolitical Narratives to the 'Administrative Area Economy'." *Eurasian Geography and Economics* 54: 57–77.

Cartier, Carolyn. 2015. "Territorial Urbanization and the Party-State in China." *Territory, Politics, Government* 3: 294–320.

Cartier, Carolyn. 2016. "The Political Economy of Rank: The Territorial Administrative Hierarchy and Leadership Mobility in Urban China. *Journal of Contemporary China* 25: 529–546.

Cartier, Carolyn. 2018. "Zone Analog: The State-Market Problematic and Territorial Economics in China." *Critical Sociology* 44: 455–470.

Cartier, Carolyn. 2019. "Exemplary Cities in China: The Capitalist Aesthetic and the Loss of Space." *Eurasian Geography and Economics* 60: 376–399.

Chan, Kam Wing. 1994. *Cities with Invisible Walls*. Hong Kong: Oxford University Press.

Chan, Kam Wing and Yanning Wei. 2019. "Two Systems in One Country: The Origin, Functions, and Mechanisms of the Rural-Urban Dual System in China." *Eurasian Geography and Economics* 60: 422–454.

Chung, Him. 2009. "The Planning of 'Villages-in-the-city' in Shenzhen, China: The Significance of the New State-led Approach." *International Planning Studies* 14: 253–273.

Douglass, Mike, Bart Wissink and Ronald van Kampen. 2012. "Enclave Urbanism in China: Consequences and Interpretation." *Urban Geography* 33: 167–182.

Faure, David and Tao Tao Liu. 2002. "Introduction." In *Town and Country in China: Identity and Perception*, edited by David Faure and Tao Tao Liu, 1–16. Basingstoke: Palgrave.

Ferencuhová, Slavomira and Michael Gentile. 2016. "Introduction: Post-socialist Cities and Urban Theory." *Eurasian Geography and Economics* 57: 483–496.

Guldin, Gregory Elyiu. 1992. *Urbanizing China*. New York: Greenwood Publishing Group.

Hamnett, Chris. 2020. "Is Chinese Urbanisation Unique?" *Urban Studies* 57: 690–700.

Harvey, David. 2005. *A Brief History of Neoliberalism*. Oxford: Oxford University Press.

He, Shenjing. 2007. "State-sponsored Gentrification under Market Transition: The Case of Shanghai." *Urban Affairs Review* 43: 171–198.

He, Shenjing and Fulong Wu. 2009. "China's Emerging Neoliberal Urbanism: Perspectives from Urban Redevelopment." *Antipode* 41: 282–304.

He, Shenjing and George C.S. Lin. 2015. "Producing and Consuming China's New Urban Space: Urban China Studies." *Urban Studies* 52: 2757–2773.

He, Shenjing and Junxi Qian. 2017. "From an Emerging Market to a Multifaceted Urban Society: Urban China Studies." *Urban Studies* 54: 827–846.

Huang, Youqin. 2020. "Introduction: New Urban Challenges in the Twenty-First Century." In *Chinese Cities in the 21st Century*, edited by Y. Huang, 1–32, Albany: University at Albany, State University of New York.

Jayne, Mark. 2018. "An Introduction to Critical Perspectives on Chinese Urbanism." In *Chinese Urbanism: Critical Perspectives*, edited by M. Jayne, 7–14. Abingdon: Routledge.

Jonas, Andrew E.G. 2020. "China's Urban Development in Context: Variegated Geographies of City-Regionalism and Managing the Territorial Politics of Urban Development." *Urban Studies* 57: 701–708.

Kirkby, Richard J. 1985. *Urbanisation in China: Town and Country in a Developing Economy, 1949–2000 AD*. London: Croom Helm.

Kong, Lily and Junxi Qian. 2019. "Knowledge Circulation in Urban Geography/Urban Studies, 1990–2010; Testing the Discourse of Anglo-American Hegemony through Publication and Citation Pattern." *Urban Studies* 56: 44–80.

Lazarus, Neil and Rashmi Varma. 2008. "Marxism and Postcolonial Studies." In *Critical Companion to Contemporary Marxism*, edited by J. Bidet and S. Kouvelakis, 309–331. Leiden: Brill.

Lebowitz, Michael A. 2012. *The Contradictions of Real Socialism: The Conductor and the Conducted*. New York: Monthly Review.

Lin, George C.S. 2007. "Chinese Urbanism in Question: State, Society, and the Reproduction of Urban Space." *Urban Geography* 28: 7–29.

Lin, George C.S. and Fangxin Yi. 2011. "Urbanization of Capital or Capitalization on Urban Land? Land Development and Local Public Finance in Urbanizing China." *Urban Geography* 32: 50–79.

Ma, Laurence and Edward W. Hanten. Co-edited. 1981. *Urban Development in Modern China*. Boulder: Westview Press.

Matusitz, Jonathan. 2010. "Collapsing the Global and the Local through Interscalar Strategy: A Glurbanization Perspective." *Planning Theory* 9: 6–27.

Nolan, Peter. 2004. *China at the Crossroads*. Cambridge: Polity Press.

Oakes, Timothy. 2019. "China's Urban Ideology: New Towns, Creation Cities, and Contested Landscape in Memory." *Eurasian Geography and Economics* 60: 400–421.

Peck, Jamie and Jun Zhang. 2013. "A Variety of Capitalism … with Chinese Characteristics?" *Journal of Economic Geography* 13: 357–396.

Qian, Junxi and Shenjing He. 2012. "Rethinking Social Power and the Right to the City Amidst China's Emerging Urbanism. *Environment and Planning A* 44: 2801–2916.

Ren, Xuefei. 2018. "From Chicago to China and India: Studying the City in the Twenty-First Century." *Annual Review of Sociology* 44: 497–513.

Shin, Hyun Bang. 2016. "Economic Transition and Speculative Urbanisation in China: Gentrification versus Dispossession." *Urban Studies* 53: 471–489.

Smith, Neil and Cindy Katz. 1993. "Grounding Metaphor: Towards a Spatialized Politics." In *Place and the Politics of Identity*, edited by Michael Keith and Steve Pile, 67–83. London: Routledge.

Solinger, Dorothy and Kam Wing Chan. 2002. "The Chinese Differences: City Studies under Socialism and Beyond." In *Understanding the City*, edited by John Eade and Chris Mele, 204–221. Oxford: Blackwell.

Sonn, Jung Won and Hyun Bang Shin. 2020. "Contextualizing Accumulation by Dispossession: The State and High-Rise Apartment Clusters in Gangnam, Seoul." *Annals of the American Association of Geographers* 11: 864–881.

Tang, Wing-Shing. 1994. "Urban Land Development under Socialism: China between 1949 and 1977." *International Journal of Urban and Regional Research* 18: 392–415.

Tang, Wing-Shing. 1997. "Urbanisation in China: A Review of Its Casual Mechanisms and Spatial Relations." *Progress in Planning* 48: 1–65.

Tang, Wing-Shing. 2006. "Planning Beijing Strategically: 'one world, one dream'?" *Town Planning Review* 77: 257–282.

Tang, Wing-Shing. 2007. "現代中国のアーバニズム/都市性一都市革命か? [Chinese Urbanism today: *Le Révolution Urbaine*?]." Translated by T. Shibata. In 都市文化理論の構築に向けて

[Toward the Construction of Urban Cultural Theory], edited by Urban Culture Research Centre, Osaka City University, 135–154. Osaka: Seibundo.

Tang, Wing-Shing. 2012. "The Pearl River Delta as 'Space of Hope': The Study on the Action Plan for the Bay Area of the Pearl River Estuary." Paper presented at 'Regional Trajectories in the Pearl River Delta: Reform and the Future in South China', China Research Centre, University of Technology, Sydney, 17–19 October, 2012.

Tang, Wing-Shing. 2014a. "Governing by the State: A Study of The Literature on Governing Chinese Mega-Cities." In *Branding Chinese Mega-Cities: Strategies, Practices and Challenges*, edited by P.O. Berg and E. Björner, 42–63. Cheltenham: Edward Elgar.

Tang, Wing-Shing. 2014b. "Where Lefebvre Meets the East: Urbanization in Hong Kong." In *Urban Revolution Now. Henri Lefebvre in Social Research and Architecture*, edited by L. Stanek, C. Schmid and Á. Moravánszky, 71–91. Surrey: Ashgate.

Tang, Wing-Shing. 2016. "Creative Industries, Public Engagement and Urban Redevelopment in Hong Kong: Cultural Regeneration as Another Dose of Isotopia?" *Cities* 56: 156–164.

Tang, Wing-Shing. 2017. "Beyond Gentrification: Hegemonic Redevelopment in Hong Kong." *International Journal of Urban and Regional Research* 41: 487–499.

Tang, Wing-Shing. 2019. "Town-country Relations in China: Back to Basics." *Eurasian Geography and Economics* 60: 455–458.

Tang, Wing-Shing. 2020. "Land Development in Hong Kong." *Equilibri* 24: 250–257. doi:10.1406/98116.

Tang, Wing-Shing. Forthcoming. "Chinese Critical Geography: A Non-Dualistic, *Tongbian*-Informed Spatial Story." In *Placing Critical Geography*, edited by L. Berg, U. Best and H. Larsen. Farnham: Ashgate.

Wissink, Bart. 2019. "Enclave Urbanism in China: A Relational Comparative View." In *Handbook on Urban Development in China*, edited by Ray Yep, June Wang and Thomas Johnson, 171–186. MA: Edward Elgar Publishing.

Wong, Kit-Ping. 2019. "Behind the Chinese Urban Dualism: Territorially-Nested Urbanization – The Case of Dongguan." *Eurasian Geography and Economics* 60: 486–509.

Wu, Fulong. 2020. "Adding New Narratives to the Urban Imagination: An Introduction to 'New Directions of Urban Studies in China." *Urban Studies* 57: 459–472.

Wu, Fulong, Fangzhu Zhang and Chris Webster. 2013. "Informality and the Development and Demolition of Urban Villages in the Chinese Peri-urban Area." *Urban Studies* 50: 1919–1934.

Wu, Fulong, Jiang Xu and Antony G.O. Yeh. 2007. *Urban Development in Post-Reform China: State, Market, and Space*. London: Routledge.

Wu, Fulong and Klaire Webber. 2004. "The Rise of 'Foreign Gated Communities" in Beijing: Between Economic Globalization and Local Institutions." *Cities* 21: 203–213.

Wu, Fulong, Shenjing He and Chris Webster. 2010. "Path Dependency and the Neighbourhood Effect: Urban Poverty in Impoverished Neighbourhoods in Chinese Cities." *Environment and Planning A* 42: 134–152.

Zhang, Li. 2011. "The Political Economy of Informal Settlements in Post-socialist China: The Case of Chengzhongcun(s)" *Geoforum* 4: 473–483.

Zhang, Li, Simon X.B. Zhao and J.P. Tian. 2003. "Self-Help Housing and Chengzhongcun in China's Urbanization." *International Journal of Urban and Regional Research* 27: 912–937.

Zhao, Yanjing and Chris Webster. 2011. "Land Dispossession and Enrichment in China's Suburban Villages." *Urban Studies* 48: 529–551.

Zhou, Yu, George C.S. Lin and Jun Zhang. 2019. "Urban China through the Lens of Neoliberalism: Is A Conceptual Twist Enough?" *Urban Studies* 56: 33–43.

Zhu, Jieming. 2002. "Urban Development under Ambiguous Property Rights: A Case of China's Transition Economy." *International Journal of Urban and Regional Research* 26: 41–57.

Exemplary cities in China: the capitalist aesthetic and the loss of space

Carolyn Cartier

ABSTRACT
Since the late twentieth century, the subject of China becoming urban has appeared in the scholarship across the disciplines while spectacular images of China's cities, and narratives about their developmental achievements, have proliferated in global media. Simultaneously, the parallel "spatial turn" in social thought invigorated geographical approaches to cities and urban change. Yet at this promising meeting ground, between contemporary geographical thought and urban-industrial transformation in China, research has tended to demonstrate a "loss of space" through patterns of dependence on analog circulation of exemplar paradigms that derive from the history of the capitalist city and liberal political economy. What drives this condition and what do we know about its practices and proliferation? This inquiry challenges research design and the politics of theory to consider how routine adoption of capital-centric concepts for research on cities in China arguably reflects the priority of paradigms in the disciplines and their conjunctures with exemplarity in Chinese society and political philosophy. These interstices facilitate application of analog models, and selective adoption of empirical information to suit them, with the paradoxical result of portraying cities in China through multiple capitalist aesthetics including a relatively narrow range of empirics framed by market-based social thought.

Introduction

In September 2011 when New Yorkers initiated "Occupy Wall Street" – the mass movement against social and economic inequality – and declared, "We are the 99%," people in Henan province in north China were among the first to respond in solidarity. In the cities of Luoyang and Zhengzhou, retirees and some local Chinese Communist Party members gathered in support of the New Yorkers demonstrating in Zuccotti Park. In Zhengzhou, they raised the banner "坚决支持美国人民伟大的华尔街革命" (Resolutely Support the American People's Great Wall Street Revolution). The short-lived response in Henan emerged in the style of slogans

This article has been republished with minor changes. These changes do not impact the academic content of the article.

and banners that characterized mass movements of the Mao era, 1949–1976 (Wu 2011). Such revolutionary sentiments in China and goals of the Occupy movement in the United States generate from significantly different histories of political and economic experience. But through space-time distanciation, and a few Internet memes, once incommensurable places even in provincial China now interrelate, juxtapose, and resonate with global events in unexpected ways.

The mediascape of China becoming urban, significant and sensational, circulates spectacles that attract worldwide interest. These include many "mosts" and "largests" such as one of the largest new buildings in the world, in Chengdu, and the most recent replica of the Eiffel Tower, in suburban Hangzhou. But the contemporary Chinese city distinctively marks the history of urban modernity because its human dimension paradoxically emerges in speed: urbanization in China is taking place not in centuries but in decades of human lifetime. People directly witness the vast scale and scope of its materialization. And yet the significance of urbanization in China, despite extensive scholarship, arguably continues to be incommensurable with understanding. Even the names of these major regional centers of millions of people – Chengdu and Hangzhou – among the most famous cities in China, elicit little recognition outside specialist scholarship. Luoyang too is only provincial in the absence of history: it was a recurring imperial capital for more than 1500 years. The reach of urban history alone in China challenges temporal baselines in comparative urban inquiry.

The potential of different histories of urban development is more than questions of culture and development, more than the always-problematic binaries of east/west, west/the rest/or west/China. They are questions about intersections and meeting grounds of ideas traveling in the world within and among disciplines and national academies, and through the ideal of the international academy – questions that lead to explanations about urbanizing China that demonstrate new insights and confidences yet simultaneously continue to occlude, bracket, and mystify what is "variegated" about China in the sense of more than variation in economic conditions as national or regional archetypes.

This discussion seeks to take the measure of possibilities and incommensurabilities in research on cities in China through questions about research design and the uses of paradigms and theory in disciplinary and interdisciplinary work. It takes a historical approach in order to locate conditions of epistemological disjuncture and to assess what constitutes their gaps and whether and in what ways they might be bridged. Its analysis builds on arguments developed by Wing Shing Tang (2014), who identifies in the literature "random conceptual appropriation" of western theories and concepts, including neoliberal urbanism, gentrification, gated community, enclave urbanism, right to the city, and so on, and their "random conceptual indigenization" through anodyne modifiers like "state-led" and passing references to government and related elocutions that, through minimal discursive modification, elide powers and agency of the party-state. By "random" I take Tang to mean adoption of concepts simply because they already exist and appear suitable, confirmed by publication in international English-language journals, and, through indigenization by selective use of empirics yet without identification of conditions that differ from those embedded in the adopted theory, and how those conditions would compel a dialectical analysis – a dialectics of theory and empirics – for consequent innovation of theoretical work. Together these conditions and their limitations confirm and reproduce the priority of disciplinary theory and, often, western theoretical baselines in history and political economy.

The discussion first situates geography in historical–global relations of scientific modernity and disciplinary paradigms and models, inclusive of China, where the development of tertiary education and university research are fundamental to the modernization project. The classic problem with scientific paradigms, following the Kuhnian thesis, is that they do not serve as incisive analytical tools but as exemplars, i.e. their explanatory elements are not evaluated for their limitations and adjusted for new situations, with the result that their applications in places whose empirics do not confirm their existing elements pivot on bracketing the different conditions or adjusting empirics to accommodate the paradigm. Analog logic, i.e. circulation and repetitive adoption of the paradigm, finds parallels of analog models in Chinese society and political-economic thought, past and present, including models for exemplary conduct and achievement. The party-state promotion of heuristic models and, through them, specification of standards in China, combined with neoliberalizing mandates for quantities of international publications, worldwide, contributes to the circulation of exemplar paradigms in scientific research, including urbanization. This intersection of models supports adoption of western *qua* international paradigms that confer on urban and regional research in China less about accounts of historical conditions and Chinese political economy than interpretations refracted through the priority of models whose reproduction accumulates in a "loss of space" about geographies in China.

By "loss of space" I mean the passing over and leaving out urban geographies of context and potential ideas about them. These contexts are more than changing local conditions or notional empirics; they are also general conditions of geographies driving the political economy that do not "fit" the normative paradigms. The "loss of space" sees this disregard of material histories and their ideational potentials spurred by adoption of formulaic exemplars. Especially in the journal literature, partial representations, and some inaccurate ones, leave China written in terms of internationalized theory – spectacular, neoliberal, and entrepreneurial. The iterative and repetitive nature of the practice of matching China to analog models yields only a refracted version of China. As Tim Oakes (2019) writes, in this issue, "urban China has come into academic and popular media focus in a way that often isolates it from China's own historical, social, and political contexts." This problem reflects in part geography's disciplinary history and traditions.

Science and geography

Across the spectrum of human and physical geography, geography demonstrates an international history of dedication to science and scientific practice (Martin 2015). Yes, theoretical advances in human geography have laid bare the limits of positivism and propelled a theoretical rejuvenation in spatial thought in the discipline (Soja 1989). Its conceptual leadership has exposed the limits of the fixed-space approaches, especially the "container" of methodological nationalism (Agnew 1987), which has plagued the social sciences at large (Wimmer and Schiller 2002). Geography has proved to be especially adroit at handling scale relations and the messy meso-scale of cities and regions, which are the contexts of economic agglomeration in the current global age. Nevertheless, it remains that the ideal and mission of science has prevailed in statements of the flagship geographical organizations across the continents.

The national geographical societies were established to pursue knowledge of the new and the legitimacy of scientific discovery – exploration, documentation, and classification. They formed in the nineteenth century when rising industrial powers sought economic frontiers and supported expeditions to the ends of the Earth. The Société de Géographie emerged first, in Paris, in 1821, followed by the Berlin Geographical Society in 1828. In Britain, the Royal Geographical Society, chartered by Queen Victoria in 1859, began as a private club in 1830. The National Geographic Society, organized in Washington, D.C. in 1888, gained popularity through dissemination of scientific maps. In debate over whether to search for the lost Arctic expedition of Sir John Franklin, which sought the Northwest Passage, elite explorers in New York formed the American Geographical Society in 1851. Only the Swedish Society for Anthropology and Geography, founded in 1877, was interdisciplinary from the start. In China, Zhang Xiangwen organized the first professional geographical organization, the Geoscience Society of China, in Tianjin, in 1909, and Zhu Kezhen led the founding of the Geographical Society of China, in Nanjing, in 1934. With establishment of the People's Republic of China (PRC) in 1949 and subsequent reorganization of academic institutions under the Chinese Communist Party (CCP), the two organizations were merged to form the current Geographical Society of China.

All these institutions, to the present, have explicitly supported geographical science. The Royal Geographical Society's "purpose remains the same today as when it was founded, namely 'the advancement of geographical science'" (Royal Geographical Society n.d.). The objective of The Geographical Society of China (2018) "aims at promoting scientific research … popularizing national geographical knowledge … making contributions to the socialist modernization drive, and developing socialist material and spiritual civilization." This expansive agenda incorporates the central mission of the PRC, the socialist modernization drive, and the core language, developing socialist material and spiritual civilization, of the party-state's general political thought program. The standard term party-state indicates in the unitary state formation that is in the PRC the power of the CCP over the government at all levels.

Consider first the language of the objectives of the Geographical Society of China. The typical international analysis elides the socialist political thought language. Spiritual civilization or *jingshen wenming* (精神文明) is a core concept in CCP thought work. Core party concepts like spiritual civilization appear in the Chinese-language journal literature, but do not, reflecting "international" paradigms, feature with any reliability in the disciplinary geography literature. Where international knowledge now identifies matters of authoritarian functionality in the PRC, like the all-state media and controls on the Internet, party-state governance of information is a more complex programmatic arena, both domestic and international, under administration of the *Xuanchuanbu* 宣传部, the Propaganda Department, also translated as the Publicity Department. Spiritual civilization is one of its umbrella concepts, featuring in party-state policy for cities, state industries, and the education system. In the 1990s, "patriotic education was formally designated as 'the main task of spiritual civilization work'" (Brady 2009, 50). Spiritual civilization work also serves as a policy program for city governance, which I return to assess in this discussion.

The contribution of geographical research to development and modernization in China extends a long history of geography in service to the state. Activity of the Royal Geographical Society was closely bound up with British colonialism (Barnett 1998; Driver 2001). French and German universities opened the first departments of geography after the Franco-Prussian War, emphasizing geographical knowledge for military preparedness (Hudson 1977). German geography developed location theories to plan settlement and industrialization in support of state strengthening, under the Third Reich (Barnes and Minca 2013). About the United States, as Neil Smith narrates, in *American Empire*, geographers worked in service to the government's geopolitical interests. From World War II, "marked by tense and intense communication between German and U.S. geographers," to the Cold War and the World Trade Organization, geographical thought in the United States informed strategic state interests in "the global geography of the American Empire" (Smith 2004, 275, 456). In American twentieth-century geopolitics, Smith observes "little critical difference between academic geography and national policy" (282).

Geography's capacity to inform national development and industrial modernization has shifted across the Pacific. The record of geographical scholarship about China makes evident the close relationship between national development and geographical research in the PRC (Fan, Ma, Pannell and Tan, 2004). Since the PRC opened to the world economy in the 1980s, the fields of economic geography, urban geography, and urban and regional planning have defined and dominated the geographical scholarship. The late industrialization of China has made its economy and urbanization the overriding subjects of domestic planning and research as well as international inquiry. On the domestic front, in addition to national research funding opportunities, province and city governments maintain funding programs for local applied research, and, similar to the history of other industrial powers, geographers in China's era of industrialization also consult for government. China's interests, capacities, and impacts have made its urbanization and economic development the normative subjects of geographical inquiry for three decades.

Research on the relationship between scientific modernization and geographical knowledge production in China, in the manner of Smith's work in *American Empire*, awaits future scholarly inquiry. We can note a few points of departure. The priority of scientific modernization compelled departments of geography in major universities to demonstrate commitment – by changing their names. There was an international petition against the change, led by Zhou Yixing, faculty member in the Department of Geography at Peking University (the conventional name of Beijing University), and Laurence J. C. Ma, in Geography at Akron University, when it became known that the university would change the name of the Department of Geography. Geography in Peking University is now Urban and Environmental Sciences. In Shanghai, East China Normal University has developed a Faculty of Earth Sciences with four departments in which three focus on the physical environment and one is the School of Urban and Regional Sciences. The national significance of urban and regional development in China makes research on cities and economic regions an expansive scientific enterprise.

The imperative of science has only intensified since 2012 under the leadership of Xi Jinping, General Secretary of the CCP and President of the PRC. His recent speeches, in 2018, for instance, continue to trumpet "scientific, technological, and industrial revolutions." As China's state news agency, Xinhua (2018), reports, at the 19th Meeting of the Academicians of the Chinese Academy of Sciences, in Beijing, "Xi urged courageous moves to reach for the 'commanding heights' in scientific and technological competition and future development." Especially the national-level universities (the highest-ranking universities, located around the country), under the direct leadership of the Ministry of Education, champion state directives and goals, and compete to attain and lead international standards.

Paradigms and politics

The challenges and complexities of research on urban China also reflect tensions between scholarly fields, especially those between the disciplines and specialist China research in area studies. At their crossroads, debate periodically surges around the relationship between theory (as if only disciplines articulate it) and empirics (as if China studies' singular utility). Contrary to typical assumptions in the disciplines, the adoption of scientific paradigms has also characterized China studies. Historically, these include applications of scientific Marxism and adaptations of spatial science models to analysis of regions in China.

My interest in the problems of paradigms in the epistemologies of knowledge emerged with the realization that research on cities and towns in China studies depended on outmoded paradigms in mid-century location theory and spatial science (Cartier 2002). Indeed, when the China field in the United States first adopted location theory for research on cities, in the 1960s, it was in response to institutional pressures for the advancement of social science. G. William Skinner, a sociological anthropologist who occupied leadership positions in the U.S. academy, developed the model of "macroregions" from central place theory and urban and regional systems theory. The model defines nine spatially bounded watershed regions in which, presumptively, separate urban systems developed and, thus, in which scholars would locate their work. In a review of the field, Farquhar and Hevia (1993, 496) characterize Skinner's "regional-market systems approach" as "a ruling organizational paradigm for social history in China." It is this role of paradigms operating sociologically as a "group-licensed way of seeing" that Thomas Kuhn (1962, 189) identified in his classic analysis, *The Structure of Scientific Revolutions*. If scholars learn the paradigm socially, e.g. their colleagues or professors insist on its importance and so on, rather than analytically, through discerning analysis, the routine adoption of the paradigm continues without questioning, and in many cases without understanding its fundamental elements.

Skinner published the model in the late 1970s, yet by this time location theory was becoming superseded in geography. From the ensuing divergence between geographical theory and China studies, the specialty field has never recovered conceptual leadership. The China field's prolonged devotion to its group optic of fixed regions arguably resulted in both limited interpretations of space and periodic returns to the foundry of geographical thought in which Skinner originally discovered the ore for his model. This is evident in the literature related to the 1990s-era spatial turn. Prominent books on urbanizing China featured new visions of space as a total concept, i.e. without warranted engagement of interrelated interpretations and ideas including landscape, place, and territory (cf. Li 2002; Yang 1999). In the wave of the poststructuralist shift in social thought, the spatial turn witnessed a variety of approaches. Some over-determined the idea of space and continued to treat space as merely bounded or fixed areas.

Even the dominant disciplinary partners of China research, namely economics, history, political science, and sociology, adopted the macroregion model. By treating river basin regions as fixed territories, the model effectively scales-down the territorial container embedded in the nation-state model. Indeed, an infrequent conceit of the core social sciences disciplines is their dependence on a naturalized nation-state paradigm (Cooper and Packard 1997; Walder 2004). The priority of bounded spaces in the geographical imagination belies the existence of dynamic spatial processes in urban and regional formation. Geography's command of ideas at the meso-scale ascended in the late twentieth century as the liberal state's Keynesian-era nation-space yielded to city-region governance in networked global economies.

Certainly, normative theory freights any disciplinary research train. But the problem is not theory *per se*; the problem is un-reflective uses of paradigms in research design. At the height of the challenge to the nation-state as a "container" of knowledge during the poststructuralist shift in the 1990s, Arif Dirlik, a historian of China's political economy, invoked the work of philosopher Paul Feyerabend to articulate a meta-argument about the relations between theory and empirics:

> Facts and theories are much more intimately connected than is admitted by the autonomy principle or 'positivist scientific practice' – the principle that facts exist and are available independently of whether or not one considers the ideological context and alternatives to the theory to be assessed. Not only is the description of every single fact dependent on some theory ... but there also exist facts which cannot be unearthed except with the help of theory, and which become unavailable as soon as such alternatives are excluded. (Feyerabend as quoted in Dirlik 1996, 246–247)

With this challenge, Dirlik identifies the problem of asserting facts and sheltering them under methodological claims of scientific practice. The issue Dirlik identifies is the adoption of research paradigms, including the macroregion, as if scientific without understanding their theoretical elements and implications – and their politics. There was a context in the American academy. As the United States industrialized to global prominence after the Second World War, not unlike China today, the social sciences came under acute pressure to demonstrate quality, merit, and utility through the adoption of scientific models (Geiger 1993; Atkinson and Blanpied 2008). In the Cold War era, liberal market ideology especially characterized the normative models. The scientific imperative marched ahead with the highpoint of postwar American industrial capitalism.

That the selection of empirical material in itself also depends on seeing through the lens of theory, acknowledged or not, advances another level of challenge. That existing sets of facts may represent particular paradigms or may be ideologically cast, imbued with bias, can pre-populate with politics what would be objective practice. Certainly, this reality may be more apparent since "alternative facts" entered the mediascape, i.e. associated with the discourse of the 45th President of the United States. Can we reliably conduct academic research with recognition of our own worldviews, and interrogate our own potential biases?

State statistical projects can be emblematic of these issues. In the 1950s, state formation in China under the People's Republic sought to classify minority groups which followed, in several ways, the approach developed in the Soviet Union (cf. Hirsch 2005). The field survey, supported by scientific work of ethnologists and linguists, ultimately resulted in official recognition of 55 groups in the PRC. Most are in large territorial areas in the west and northern and southern border regions, but they constitute only a fraction of the more than 400 ethnolinguistic groups identified in the research work (Mullaney 2010). Neither the diversity of national minorities nor the stability of their cultures and livelihoods would be the scientific priority in advancing territorial governance.

Exemplar paradigms

In contrast to the disciplines of history, economics, political science, and sociology, geography and anthropology have attained a certain breakthrough status in research on China because their epistemological advances grapple with dynamics between and among local, national, and global conditions. In lines of inquiry that trace from David Harvey's (1978) early theoretical work on the urban process under capitalism right up to the present, urban and economic geography has prominently demonstrated capacity to frame questions about precisely the predominant changes at stake in contemporary China – urbanization and development of cities and regions. Fulong Wu's first article in the *International Journal of Urban and Regional Research*, for example, is an interpretation of property rights and Harvey's thesis of the urban process for China (Wu 1997). Urbanization in China underpins the country's domestic growth and developmental modernization. Yet paradigms in urban geography are precisely those whose appropriation Tang has cautioned us to assess. This is not a matter of internecine politics in the discipline or urban studies: in the humanities and social sciences at large, since the spatial turn, multiple disciplines and various interdisciplinary projects have routinely idea-borrowed from currents in geographical thought.

How paradigms gain institutionalization in research design depends on the mutually constitutive process of adoption and circulation among knowledge communities. Kuhn's treatment of scientific practice and scientific breakthroughs demonstrates the significance of the social in the pursuit of science. Kuhn observes how paradigms circulate through scholarly networks that assiduously adopt and promote particular models and approaches. Contrary to principles of scientific practice, Kuhn finds, research networks reproduce paradigmatic models more than they actually test their elements. These networks and groups may not even be likeminded, yet contend over the same model. Developing new ones is typically not the priority.

We see this situation in the journal literature, in which academics debate the details of each other's work. In one memorable case, Neil Brenner, focused on the primacy of capitalism in dynamic rescaling, importunely critiqued the dimension of household economy that Sally Marston brought to the subject. Mark Purcell coined the phrase "island of practice" to characterize the Marston/Brenner debate because, instead of advancing the scholarship, it produced "nonengagements." Purcell writes, identifying the problem of discrete paradigmatic worldviews, such islands "close down analysis so that scholars have difficulty seeing past their particular focus" (Purcell 2003, 328). He urges the development of a synthetic geography that would collaborate across research subfields. Kuhn, in reflection on the general issues and longer-term implications for the quality of knowledge production, developed the thesis that scientific breakthroughs more often appear in the work of relatively independent scholars who pursue original projects.

In response to his original thesis, Kuhn (1970) subsequently defined two kinds of paradigms, the "exemplar" and the "disciplinary matrix." The matrix constitutes the broader knowledge of the discipline whereas the exemplars are specific models. The particular problematic for scholarship on urban China is the circulation of exemplar paradigms. In practice, it works like this: an exemplar paradigm modeled on the history of cities in western Europe or the United States, for instance, circulates through scholarly adoption and publication to model places with histories that are not based on historical political-economic liberalism. The exemplar paradigm does its incremental analog work by masking the crucial issue of consequential differences between place of origin and the place of application. The disciplinary *coup de grâce* is the chorus of complaint that differences in the political economy of urban institutions do not really matter if we are interested in the realities of general change in the contemporary world economy. Alan Scott's work on the city-region, for example, based on economic and population agglomeration, demonstrates this tendency (e.g. Scott 2019).

The problem of analog circulation is by no means new. Indeed, it constituted a short-lived explicit research approach. Andrew Mair, a student of Kevin Cox, spotlights the history. In dialog with Kuhn, Mair identifies how the "exemplar method," transmuting "a *pedagogic* as well as a *heuristic* role," uses "solutions to past problems in order to account for new situations" but "without recourse to explicit correspondence rules or even empirical generalizations" (Mair 1986, 351). He continues, "a subset of analogical arguments is uniquely geographic; the application of analyses derived in one place to new situations in other places" (360). For instance, Mair identifies in Michael Woldenberg's and Brian Berry's (1967) attempt to compare river systems and central place systems a "stillborn" effort with "little positive content." Woldenberg sought to model "persistent geometric progressions" of "mathematical regularities" in physical, human, and biological forms. In addition to attempting to naturalize market economies, i.e. comparison of central place systems and river systems, Woldenberg (1979, 429) proposed comparison of "trees, blood vessels, airways, bile ducts and even the microscopic branching of the Purkinje cell in the brain." The faulty logic on display is the one in which would-be similar patterns are held to be significant and explicable by mathematical generalization. The analog approach seeks to identify regularities of similar order – similar forms, patterns, structures, and internal arrangements. Self-evident problems – great leaps among subjects and domains characterized by widely if not wildly different conditions – confirm the unsound analog method.

Such illogic also appeared in the field of spatial science in the postwar era. In deservedly limited research on "social physics," the laws of physics were applied to social change. Trevor Barnes explains the analog: based on Newton's gravity model, researchers "assumed that humans interact over space as heavenly bodies do in the celestial system" (Barnes 1994, 567). There is the laughable element, but the fundamental scientific problem is faulty analog reasoning.

In more recent work, Giorgio Agamben (2009) too observes how analog circulation limits the quality of exemplar models in research design. Instead of working to inform deductive and inductive reasoning between the general and particular, the *sine qua non* of geographical research, the exemplar paradigm as analog becomes, through iterative repetition, a singularity that fixes paradigmatic meaning. Regular analog exposition constitutes its intelligibility: a model becomes popular because it has become well known, not because it is necessarily epistemologically sound, scientifically tested, or adjusted and changed in relation to new discoveries.

In urban geography and comparative urban studies, the concept of gentrification exhibits analog circulation. Until just over a decade ago, gentrification research focused on the U.K. and North America, with disproportionate work on London and New York. Since then, the concept has spread while generating debate over its meanings, utility, and effectiveness. Among other issues, it stretches awkwardly to grasp realities where the state takes a leading role in the redevelopment of the city. Scholars of China, Hong Kong, and Singapore, for example, find its conceptual elements lack the capacity to handle structural conditions of historical land development (Tang 2017) in addition to the multiple roles of the state in the redevelopment of urban housing (Haila 2017; Tomba 2017).

Debate over the gentrification concept also extends to analog plays in the international academy. For example, scholars of gentrification "call for" gentrification research to be extended globally (Lees 2012; Ley and Teo 2014). On one hand, such efforts represent concern for serious urban issues as if neglected. On the other, they can suggest academic hegemony as if research on urban redevelopment, class, and inequality is lacking "elsewhere." In my analysis of the issue, it became apparent that David Ley, who suggested gentrification was effectively "missing" in Hong Kong, had both overlooked existing scholarship while overstating the appearance of gentrification in various national conversations (Cartier 2017).

Promotion of the English-language word gentrification, globally speaking, resonates with disciplinary theory's disappearance of inconvenient empirics. Attempts to promote gentrification exhibit analog circulation, as well as invite postcolonial critique of the apparent paradox: "western" or "northern" scholars claim some "lack" elsewhere while lacking knowledge of existing research in "other" academies. Existing postcolonial critique calls this move epistemic violence. I would like to draw on Chinese-language phraseology and call it the "double-lack." Indeed, theoretical advances in geography, leading the spatial turn, also addressed such inequities in the international academy. We have to be vigilant. My main concern about this kind of analog borrowing is its capacity to occlude from research inquiry actual causal processes at stake.

The problem of analog properties for geographical exemplars is their application to apparently "like" places, places that the exemplar was not developed to theorize. Consider the "global city" as a paradigm, based on the work of Saskia Sassen for London, New York, and Tokyo. A certain amount of debate has demonstrated the limitations of treating Tokyo as a global city by comparison with London and New York (White 1998, 2011). The comparative scholarship on China has predictably sought to examine Shanghai as a global city. A relatively new financial district of super-tall buildings commands Shanghai's urban landscape, generating the appearances of a global city. But the substantial scholarship on the structural conditions of the Shanghai economy finds them to be sufficiently different from those of London and New York to conclude that Shanghai does not share diagnostic global city conditions (Li and Wu. 2006). The central government of China directly administers Shanghai, whose conditions differ in key ways from assumptions about capitalist rescaling and global networks (Cartier 2015). The global city concept gained altitude through worlding discourses, while the realities of some major cities are only refracted through its lens.

The problems of the exemplar for urban geographies in China can be resolved only partially through critique of theory. Varieties of urban theory, whether neoclassical (spatial science), Marxist (the urban process), or neoliberal (politics of scale), all based on the capitalist city in Europe and North America, tend not to incorporate geographies of power. As John Allen (2003) has demonstrated, varieties of power tend to be "lost" in capital-centric thought. Resting on the history of political-economic liberalism, varieties of urban theory do not reliably bring into view the absences they otherwise generate, including the many powerful roles of the party-state governing urbanization in China. Urban theory tends to reproduce the capitalist aesthetic – ideas about and materializations of the circulation of capital. And yet we cannot simply call this a problem of western theory because research on urban transformation in the Chinese academy has also hewn to normative international paradigms. If geographers and scholars in the Chinese academy have also adopted so-called western theory, then the problem is not the West perforce; the problem is inextricably bound-up with the expansion and internationalization of academic work in and about China and Asia at large.

One line of response appears in the analysis developed by Chan and Wei (2019), in this issue, on the origins of the Mao-era rural-urban dual system in the Soviet Union. The Soviet Union first developed the rural-urban dual system in which agricultural productivity was siphoned off to support urban-industrial development through collectivization, state procurement, and limits on rural-urban migration. This dual system organization transferred to China during the 1950s when the Soviet Union advised the PRC. The *hukou* system and the land use system in China today derive from the Soviet model. However, comparative analysis of this significant history appears infrequently in scholarship. The article by Chan and Wei is the first in a geography journal to analyze the political-economic origins of the Chinese socialist economy in the Soviet system. This reality contributes another paradoxical question to our concerns: why do paradigms adopted for research on China derive more often from the experience of the urban process under capitalism rather than its legacies and transformations from the history of the planned economy under socialism?

Models and modeling achievement

In late 2012 the international news media ricocheted with reports of the largest building in the world by floor area about to be opened in Chengdu, the next in a long line of "largests" in China's cities. Thomas Campanella, in *The Concrete Dragon*, chronicles the "sheer audacity" of it all in terms of Schumpeter's creative destruction: "Bigness and supersized sprawl may have once been American specialities, but that monopoly has been usurped" (Campanella 2008, 14–15). The words market and capitalism feature liberally in his narrative, but socialism appears as only a vestige of the past. Campanella strikes the common equation between the spectacular built environment in China, cities, and capitalism.

Yet the distinctiveness of the Chengdu project resonates with Tao Zhu's diagnostic of the socialist built environment in China since the 1950s – "building big, with no regret" (Zhu 2011). In 1958–59, in a little over a year, the Party and the central government led the construction of the Ten Great Buildings in Beijing, which included the Great Hall of the People, Beijing Railway Station, and the National Museum of China to commemorate the 10th anniversary of the People's Republic of China. In recent years, rapid construction of large-scale built environments, including infrastructure systems, has become as if diagnostic of China's economic growth in recent decades. But the scale of the projects continues the history of state-led construction under socialism.

Direct comparisons exist in the history of Soviet planning. In the Soviet era, massive infrastructure construction was called "total planning" for simultaneous planning of the economy and society. The Soviet leadership, Collier (2011, 38) recalls, sought rapid industrialization through "city-building" or *gradostroitel'stvo*, which was not limited to the construction of new cities. It was the general program for coordinating the populations and the political economy for industrial modernization. It sought to "revolutionize the national population's conditions of existence, first of all by rearranging it over national space" (49). This is the city-space of the socialist state's *telos* of planning – the space economy of the state's idealized future in urban form. This role of the city in total planning has re-emerged in China, not only for urbanization but also scientific modernization and general social and economic development.

In work on the built environment, Bianca Bosker (2013) asks why so many urban development projects with self-consciously iconic buildings reproduce western architectural forms. The seven versions of the Eiffel Tower in China are only some of the most recognizable among imitation structures and places, which include numbers of White Houses, Venices, and Versailleses in addition to Disneylands. In analog circulation, based on precepts of the exemplary model, any one of the replica Eiffel Towers in China may be read less as a kitschy copy than a model of iconicity. Bosker draws on Chinese aesthetics to venture that architectural copies should be understood in relation to Chinese art and imperial microcosms. In Chinese painting, for example, good imitation of an existing work, demonstrating inspired interpretation, is to be well regarded. Thus, what is the local quality of the Tower Bridge replica in Suzhou? Is it a good rendition, well built and the well situated? Bosker also considers the role of imperial microcosms to resituate interpretations of space and place. China's imperial gardens, stocked with representative flora and fauna from across the empire, constituted microcosms of imperial rule. Wealthy merchants built their own, in a manner of landscape appropriation, which symbolized legitimacy. After the PRC opened to the world economy, property developers built replicas of iconic foreign places that the new Chinese consumer-citizen, now imagining globally, would seek to experience. This arguably closer relationship between reality and representation in Chinese aesthetics, including architecture and landscape design, influences perceptions of places – making copy architecture another exercise in space-time distanciation.

Exemplarity at large

Exemplary models have a long history in China, but the notion of the CCP promoting didactic models for citizen socialization tends to be associated with the Mao era. However, the Party has revived didactic models in recent years in association with scientific development, including models of exemplary human conduct and places of exemplary development. Through the circulation of exemplary models, now in the name of science, the party-state conveys of what it approves and seeks to socialize commensurate conduct.

In *The Exemplary Society*, Børge Bakken examines the revival of models, called *moshi* 模式or *mofan* 模范, of people and places and moral models or *daode mofan* 道德模范 in relation to scientific development and modernization. Exemplary models constitute elements of "the 'scientific' approach to human and social improvement," writes Bakken (2000, 30). Exemplary persons range from contemporary leadership figures to Mao-era socialist avatars and modern representations of Confucius. In recent decades, the "models should reflect modern values, with the fixity of the traditional exemplary model loosened up in order to adapt to the changing times." What has changed is the "scientification of exemplarity, for the production of norms and standards to be carried out in a far more systematic and scientific manner" (Bakken 2000,

217). In turn, the scientification becomes institutionalized through "a methodology of spiritual civilization based on the scientific production of norms and standards." Here we see the meta-concept of civilization, which appears uniformly in governing institutions, including city governments, schools, and workplaces. The party-state has renovated the significance of models for precepts of modern scientific development.

Behavioral models trace to the tradition of scholar-officials as exemplars of learning and leadership (Munro 2000), yet the compelling matter is how and why these models continue and transform. In the 1940s, the CCP introduced the Stakhanovite labor mobilization model from the Soviet Union to promote zealous work activity (Funari and Mees 2013), and subsequently portrayed the urgency of socialist production through model places for agricultural and industrial development, uniting workers at the grassroots (Shapiro 2001).

Since the 2000s, models of both people and places have been revamped for the urbanization drive. The scientific model of human transformation appears in a biopolitics of human "quality" or *suzhi* 素质 that would modernize populations for rapid urbanization (Anagnost 2004). The party-state also seeks the people's urgent cooperation for consumer capitalism, even reviving the visage of Lei Feng, Mao's hagiographical loyal soldier and model super-citizen – representing selfless relations between the people, the CCP, and the People's Liberation Army – for placement on personal consumption items. Lei Feng may be the most obvious symbol of party propaganda, but the subject of Lei Feng features in more than 11,000 journal articles in the Chinese academy since 1963. "In recent decades," Link (2015, 27–28) writes, "Chinese Communist Party leaders have tried to revive the traditional moral-political model with certain modern adaptations" based on attaining the "Chinese dream" of wealth and power through "a 'Chinese model' of development." The renewed approach to exemplarity bridges party leadership with national development to hold out the ideal of a better life for the people, albeit now relentlessly based on consumer modernity.

In the second half of the 1990s, the improvement of urban governance for cities in China emerged in the form of model city award programs. The National Environmental Protection Model City award program started in 1997, followed by model city awards for everything from managing public security and quality of general hygiene, to innovation in science and technology and more. The pre-eminent model program, the National Civilized City, recognizes general urban governance based on measurable standards in multiple categories. It combines party priorities for official rectitude, which have intensified since 2012, with measurable governing targets, including environmental quality, the provision of cultural facilities, and food security. But its origins lie in spiritual civilization thought work. In the late 1990s, the Party established the Spiritual Civilization Committee, under its Propaganda Department, also known by its official contemporary translation, the Publicity

Department, which established and continues to administer the National Civilized City award program (Cartier 2013). The program links model conduct of party officials with urban improvement, making the honorific title contingent on exemplary leadership. It is, in a Foucauldian vocabulary, a technology of governmentality for coordinated governance. Its objective measures of assessment would raise scientific management over political favoritism. Continuing to hold the title requires periodic reassessment, entraining the city in a future-oriented programmatic of striving to improve. Commensurate with PRC's interest to enter world rankings on all fronts, from world-leading productive capacity to world-class cities and universities, the National Civilized City award projects a future nation of uniformly "civilized" places.

Zone analog

Perhaps the most successful analog in the history of the reform era is the simple concept of zone and its debut as the Shenzhen "special economic zone." The PRC opened to foreign direct investment through Shenzhen, on the border of Hong Kong, placing an anchor on the edge of the harbor of global capitalism. As Hong Kong capital and manufacturing expertise crossed to Shenzhen, seeking lower costs of production, the new industrial economy grew quickly. And so international economic interest trained attention on Shenzhen, generating new narratives about "miracle" zones. In the process, the notion of a special economic zone became a space of capital circulation, a set of assumptions about economic growth, and a would-be policy approach.

Yet despite the inherent spatiality at stake, the scholarship on zones has often treated their geographies as once established and fixed, and then as sites where labor and capital jostle for product and accumulation (Cartier 2018). Aihwa Ong sought to retrieve their dynamics through "zoning" as "spaces of political and economic exception" (Ong 2004, 70; cf. 2006, *passim*). The suggestion that these areas exist as if apart from state territory facilitated the notion of "zone as exception" as an emerging paradigm. But a zone in China is only a selective translation of different types of administrative divisions. In China's system of administrative divisions, in Chinese, *xingzheng quhua tixi* 行政区划体系, the third character, *qu*区, means area, district, region or zone. The full sweep of China's subnational territory exists within this system. No areas are excepted from its system of territorial governance, just as there are no unincorporated areas.

The primary institutional category of a special economic zone, in Chinese *jingji tequ* 经济特区 or "economic special district," is an administrative division of Chinese state territory. But the notion of exception would lead Ong to portray these administrative areas as if autonomous. "In short, the coastal zone authorities and open cities are spaces of exception to the centrally planned socialist economy. They enjoy autonomy in all economic and administrative matters" (Ong 2004, 81, 2006, 108). This notion of autonomy conveys the zone's capitalist aesthetic. Here Ong conflates opportunities for workers and firms with the administrative ontology and governing institutions of state territory. In reality, Shenzhen was first established as a sub-provincial-level city, and then given designated economic functions by the Chinese central government. Like any other subnational territory in the PRC, so-called zones are territorial governing areas under the jurisdiction of and directly governed under the authority of party committees. By institutional definition, none have autonomy.

Nevertheless, propelled by analog logic, the idea of special zones as if solutions to economic problems began to travel. The economist Barry Naughton identified the issue. Barry (2009, 3) asks, "What Chinese institutions can we recommend for adoption by other developing or transitional economies?" "None," yet "we must deal with one apparent exception. Hasn't the Chinese policy of Special Economic Zones (SEZs) been replicated around the world?" "What is most innovative about Chinese SEZs is size, multi-functional use as a laboratory of multiple types of reforms, and openness to the domestic economy" which makes them actually "not a good candidate for replication." Here Naughton identifies the analog problem. Instantiated within the PRC territorial system, the SEZ is not a good candidate for international replication since it is a party-state project with state-allocated budget and exceptional powers of authority, in addition to area size on the scale of entire cities and regions. When India, for instance, sought to model China's zones, it realized their scale and scope could not be matched (Sharma 2015).

The local differences of regional development have also been expressed in the language of models. Substantial discussion takes place in China over the various merits of economic development in the provinces (e.g. Xin and Liu 2001; Linfei 2009). Analysts of regional development identify and debate some ten different models, while three major models have dominated the conversation: the Su'nan, Zhujiang, and Wenzhou models. Here the word model is also commensurable with the relatively routine discourse of international political economy. Yet its valences are more complex, linking to the broader cultural-historical discourse of exemplary practices in China.

During 2010–2012 a new discourse of regional models emerged in a competitive dialog between the party leaders of Guangdong and Chongqing (Cartier and Tomba 2012). Splashed across the media, this "cake debate" parsed comparative merits of the size, portions, and distribution of economic resources (in China, cake is better than pie). The Guangdong model would reenergize marketization for new rounds of growth, whereas the Chongqing model would continue state-owned enterprises and emphasize redistributive measures including social housing. But the popular and policy discourses of the Chongqing model began to fade, after March 2012, when the party leader of Chongqing was removed from power. To oversee the political transition, the central government assigned

Zhang Dejiang, a former governor of Guangdong province, to Chongqing. In response to questions at the 18th Party Congress, Dejiang. (2012) stated, predictably, "I think fundamentally there is no Chongqing model." But issues at the heart of the cake debate characterize ongoing issues of socio-economic inequality in China.

The "loss of space"

Exemplary models continue to circulate in China. Some play significant roles in the urban process. Others, like Lei Feng, are targets for jokes. Nevertheless, from historical personages, to contemporary party policy programs and expressions of regional development, the multiple forms of the exemplary model testify to their prevalence. These models, like paradigms, circulate through information networks, and are by definition forms of both ideology and practice. They materialize in places and some, like the awards for good governance, influence the transformation of the city. They are party-defined techniques of governmentality, working to show the acceptable ontological range for people and places under unitary state formation. Characteristic of and transcending the socialist era, models in China's culture and politics are also deeply historic in form and substance. Even as disciplinary paradigms do not facilitate research on these exemplars, their existence and analog circulation point paradoxically to the general problem of analog models in the international academy. And the word model itself, at once common and scientific, camouflages the intermediating political work the analog accomplishes, not only in China.

If exemplary models in China foreground contexts of party-state approval, what do they leave out? In general, the exemplar paradigms for urban research lack capacity to conceptualize the state structure of the urban process in China. Acknowledgment of this realization has taken a long time to emerge. A main reason why is that China's opening to the world economy converged with late capitalism's trajectory of neoliberalization, which co-generated widespread assumptions about decentralization and marketization. The neoliberal critique, circulating as a kind of free radical of theory, also attached to urban China geography. But it has lacked the capacity, for instance, to conceptualize party-state power relations in China's urban development. Indeed, with concern about the limits of the neoliberal thesis for understanding cities in Russia, Collier (2011) turned to examine the continuities of Soviet planning history. In addition, the scale relations paradigm, generated from work on state rescaling under-advanced capitalism, has also fallen short, not least because its elements do not conceptualize rank relations of the Chinese state structure. In an article on the recentralization of power over city-regions in China, Wu (2016, 1151) rehearses the international rescaling approach only to ultimately conclude "Chinese state dominance in governance is distinctively different from

Western and other non-Western contexts," yet without causal or historical analysis. Here is the realization that Wing Shing Tang identifies and articulates.

Conjunctures emerge among different kinds of models and standards. Striving for city recognition in China, from outsize construction projects, to economic productivity and urban quality awards, also appears prominently in China's education sector. Since the 2000s, the PRC Ministry of Education has worked to vault the leading universities into the higher echelon of world rankings (Wang, Wang, and Liu 2011; Ministry of Finance 2012). Within the university, the institutional remit incentivizes research productivity, including requirements for international English-language journal publications. In the process, the Chinese academy has directly linked its universities to the peer-review journal process, driving the conjuncture between the Chinese academy and the international academy. This process contributes to intensifying the circulation of exemplar paradigms.

Now the consequences become clearer. The top-down directive for scientific research pressures adoption of international paradigms. The use of exemplar paradigms for journal publication, entrained in national scientific development, becomes routine. The practice contributes to national legitimacy, even as it works to bracket original approaches in addition to existing realities or empirical conditions that do not "fit" the models. In this dialectic, the circulation of exemplar paradigms comes to serve the state, by simultaneously enhancing international legitimacy, yet through the use of models that do not investigate it. The sum total of these operations diminishes the potential for original work and the insights it would generate. We have a concentration of research approaches that has been limiting understandings of the urban process in China – the annihilation of "space" by analog circulation of exemplar paradigms.

International paradigms in urban studies, based on fundamentals of economic liberalism, typically background the role of the state. For research on China, this reasonably well-known problem has been attributed to the notion of "sensitivity" about the political system, generating some compliant ignorance about it. However, this so-called black box, as if containing the party-state from inquiry, has become a false proposition. The CCP itself says so. In 2015, the Chinese state media announced "Chinese Communist Party publishes 'self-decoding' series for Western audience." This Xinhua (2015) headline describes the CCP's "move to reach out to the Western audience to help them understand why it remains the single, largest ruling party" through a series of books. The five English-language volumes, launched in New York City at a major publishing fair, constitute the Party's self-introduction to the world.

Analysis of multiple interrelated conditions about the uses of paradigms, in pursuit of scientific legitimacy, demonstrates why research on urban China has depended on exemplars, and how their uses limit ways of seeing. Their widespread adoption has precipitated a method of matching data and information.

Neoliberalization of the academy, with normative science linked to academic promotion and national standing in the international academy, has intensified the practice. Even would-be expert research paradoxically disregards information and knowledge that is not commensurable with prevailing paradigms. These issues, characteristic of geography's scientific genealogy and the roles of science in industrial modernization, are not new. But they have arguably expanded in scale and scope. The consequential "loss of space" emerges just when China's extraordinary transformation wants for substantial documentation. At best, like the zone exemplar, the priority of market-based paradigms has been telling the story of the state only in terms of the market. Can we restore a fuller appreciation of urban China?

Acknowledgments

Funding from the Australian Research Council Discovery Projects (DP120101901, DP170100871) made possible the research for portions of this work. I owe particular thanks to Kam Wing Chan, Tim Oakes, and Wing Shing Tang for discerning suggestions on drafts of the paper. I also thank Michael Somers for reading and editing the manuscript and urging further development of its ideas. My work on paradigms and inter/disciplinary history reflects earlier intellectual support from three key scholars. The imprimatur of the historian Frederic Wakeman for reevaluation of the macroregion model mattered pivotally. Enthusiastic regard for the geographical critique from Neil Smith lent a significant dose of predictably undaunted resolve. And Geremie Barmé's shared grasp of civilizational thought contributed to ideas about reassembling the city.

Disclosure statement

No potential conflict of interest was reported by the author.

Funding

This work was supported by the Australian Research Council [DP120101901]; Australian Research Council [DP120101901; DP170100871];

References

Agamben, Giorgio. 2009. *The Signature of All Things: On Method.* New York: Zone Books.

Agnew, John. 1987. *Place and Politics: The Geographical Mediation of State and Society.* Boston: Allen & Unwin.

Allen, John. 2003. *Lost Geographies of Power.* Oxford: Blackwell.

Anagnost, Ann. 2004. "The Corporeal Politics of Quality (Suzhi)." *Public Culture* 16 (2): 189–208. doi:10.1215/08992363-16-2-189.

Atkinson, Richard, C., and William. A. Blanpied. 2008. "Research Universities: Core of the US Science and Technology System." *Technology and Society* 30 (1): 30–48. doi:10.1016/j.techsoc.2007.10.004.

Bakken, Børge. 2000. *The Exemplary Society: Human Improvement, Social Control and the Dangers of Modernity in China*. Oxford: Oxford University Press.

Barnes, Trevor. J. 1994. "Social Physics." In *The Dictionary of Human Geography*, edited by Ron. Johnston, Derek. Gregory, and D. M. Smith, 567. Oxford: Blackwell.

Barnes, Trevor. J. 2013. "Nazi Spatial Theory: The Dark Geographies of Carl Schmitt and Walter Christaller." *Annals of The Association of American Geographers* 103 (3): 669–687. doi:10.1080/00045608.2011.653732.

Barnett, Clive. 1998. "Impure and Worldly Geography: The Africanist Discourse of the Royal Geographical Society, 1831–73." *Transactions of the Institute of British Geography NS* 23: 239–251. doi:10.1111/j.0020-2754.1998.00239.x.

Barry, Naughton. 2009. "Singularity and Replicability in China's Developmental Experience." *China Analysis* 68: 1–22.

Bosker, Bianca. 2013. *Original Copies: Architectural Mimicry in Contemporary China*. Honolulu: University of Hawaii Press.

Brady, Anne-Marie. 2009. *Marketing Dictatorship: Propaganda and though Work in Contemporary China*. Landham, MD: Rowman & Littlefield.

Campanella, Thomas. 2008. *The Concrete Dragon: China's Urban Revolution and What It Means for the World*. Princeton, NJ: Princeton University Press.

Cartier, Carolyn. 2002. "Origins and Evolution of a Geographical Idea: The Macoregion in China." *Modern China* 28 (1): 79–112. doi:10.1177/009770040202800103.

Cartier, Carolyn. 2013. "Building Civilised Cities." In *China Story Yearbook 2013: Civilising China*, edited by G. R. Barmé and J. Goldkorn, 256–285. Canberra: Australian Centre for China in the World.

Cartier, Carolyn. 2015. "Territorial Urbanization and the Party-state in China." *Territory, Politics, Governance* 3 (3): 294–320. doi:10.1080/21622671.2015.1005125.

Cartier, Carolyn. 2017. "Contextual Urban Theory and the 'appeal' of Gentrification: Lost in Transposition?" *International Journal of Urban and Regional Research* 41 (3): 466–477. doi:10.1111/1468-2427.12498.

Cartier, Carolyn. 2018. "Zone Analog: The State-Market Problematic and Territorial Economies in China." *Critical Sociology* 44 (3): 455–470. doi:10.1177/0896920517712367.

Cartier, Carolyn, and Luigi, Tomba. 2012. "Symbolic Cities and the 'Cake Debate'." In *China Story Yearbook 2012: Red Rising, Red Eclipse*, edited by G. Barmé with J. Goldkorn, C. Cartier, and G. Davies, 25–52. Canberra: Centre for China in the World.

Chan, Kam, Wing, and Yanning Wei. 2019. *Two Systems in One Country: The Origin, Functions, and Mechanisms of the Rural-Urban Dual System in China. Eurasian Geogaphy and Economics* 60 (4).

Cindy Fan, C., Laurence J. C. Ma, Clifton W. Pannell, and K. C. Tan. 2004. "Geography of China." In *Geography in America at the Dawn of the 21st Century*, edited by Gary. L. Gaile and Cort. J. Willmott, 670–680. New York: Oxford University Press

Collier, Stephen J. 2011. *Soviet Social Modernity: Neoliberalism, Social Modernity, Biopolitics*. Princeton, NJ: Princeton University Press.

Cooper, Frederick, and Randall Packard, eds. 1997. *International Development and the Social Sciences: Essays on the History and Politics of Knowledge*. Berkeley, CA: University of California Press.

Dejiang, Zhang. 2012. "Zhang Dejiang: Wo Renwei Genben Jiu Bu Cunzai Chongqing Moshi" (Zhang Dejiang: I Think Fundamentally There Is No Chongqing Model)" (*Caixin*, 8 November, 2012. Accessed 8 August 2019. http://china.caixin.com/2012-11-08/100458064.html

Dirlik, Arif. 1996. "Reversals, Ironies, Hegemonies: Notes on the Contemporary Historiography of Modern China." *Modern China* 22 (3): 243–284. doi:10.1177/009770049602200301.

Driver, Felix. 2001. *Militant Colonialism: Cultures of Exploration and Empire*. Oxford: Blackwell.

Farquhar, Judith B., and James L. Hevia. 1993. "Culture and Postwar American Historiography of China." *positions* 1 (2): 486–525. doi:10.1215/10679847-1-2-486.

Funari, Rachel, and Bernard Mees. 2013. "Socialist Emulation in China: Worker Heroes Yesterday and Today." *Labor History* 54 (3): 240–255. doi:10.1080/0023656X.2013.804269.

Geiger, Roger L. 1993. *Research and Relevant Knowledge: American Research Universities since World War II*. New York, NY: Oxford University Press.

Haila, Ann. 2017. "Institutionalization of the 'Property Mind'." *International Journal of Urban and Regional Research* 41 (3): 500–507. doi:10.1111/1468-2427.12495.

Harvey, David. 1978. "The Urban Process under Capitalism: A Framework for Analysis." *The International Journal of Urban and Regional Research* 2 (1–3): 101–131. doi:10.1111/ijur.1978.2.issue-1-3.

Hirsch, Francine. 2005. *Empire of Nations: Ethnographic Knowledge and the Making of the Soviet Union*. Ithaca, NY: Cornell University Press.

Hudson, Brian. 1977. "The New Geography and the New Imperialism: 1870–1918." *Antipode* 9 (2): 12–19. doi:10.1111/anti.1977.9.issue-2.

Kuhn, Thomas. 1962. *The Structure of Scientific Revolutions*. Chicago, IL: University of Chicago Press.

Kuhn, Thomas. 1970. *The Structure of Scientific Revolutions*. 2nd ed. Chicago, IL: Chicago University Press

Lees, Loretta. 2012. "The Geography of Gentrification: Thinking through Comparative Urbanism." *Progress in Human Geography* 36 (2): 155–171. doi:10.1177/0309132511412998.

Ley, David, and Sin Yih Teo. 2014. "Gentrification in Hong Kong? Epistemology Vs. Ontology." *International Journal of Urban and Regional Research* 38 (4): 1286–1303. doi:10.1111/1468-2427.12109.

Li, Zhang. 2002. *Strangers in the City: Reconfigurations of Space, Power and Social Networks within China's Floating Population*. Stanford, CA: Stanford University Press.

Li, Zhigang, and Fulong Wu. 2006. "Socioeconomic Transformations in Shanghai (1990--2000): Policy Impacts in Global–National–Local Contexts." *Cities* 23 (4): 250–268. doi:10.1016/j.cities.2006.01.002.

Linfei, Song. 2009. "Zhongguo 'Sanda Moshi' De Chuanxin Yu Molai (the Innovation and Future of China's 'three Major Models')." *Nanjing Shehui Kexue (Nanjing Social Sciences)* 2009 (1): 1–6.

Link, Perry. 2015. "What It Means to Be Chinese." *Foreign Affairs* 94 (1): 25–31.

Mair, Andrew. 1986. "Thomas Kuhn and Understanding Geography." *Progress in Human Geography* 10 (3): 345–69. doi:10.1177/030913258601000302.

Martin, Geoffrey J. 2015. *American Geography and Geographers: Toward Geographical Science*. Oxford: Oxford University Press.

Ministry of Finance. 2012. "Report on the Implementation of Central and Local Budgets for 2011 and on Draft Central and Local Budgets for 2012." Fifth Session of the Eleventh National People's Congress, 5 March 2012. Accessed 8 August 2019. http://online.wsj.com/public/resources/documents/2012NPC_FinanceWorkReport_English.pdf

Mullaney, Thomas S. 2010. *Coming to Terms with the Nation: Ethnic Classification in Modern China*. Berkeley, CA: University of California Press.

Munro, Donald J. 2000. *The Concept of Man in Contemporary China*. Ann Arbor, MI: Center for Chinese Studies, University of Michigan.

Oakes, Tim. 2019. "China's Urban Ideology: New Towns. Creation Cities, and Contested Landscapes of Memory." *Eurasian Geography and Economics* 60 (4).
Ong, Aihwa. 2004. "The Chinese Axis: Zoning Technologies and Variegated Sovereignty." *Journal of East Asian Studies* 4 (1): 69–96. doi:10.1017/S1598240800004392.
Ong, Aihwa. 2006. *Neoliberalism as Exception: Mutations in Citizenship and Sovereignty.* Durham, NC: Duke University Press.
Purcell, Mark. 2003. "Islands of Practice and the Marston/Brenner Debate: Toward a More Synthetic Critical Human Geography." *Progress in Human Geography* 27 (3): 317–332. doi:10.1191/0309132503ph436oa.
Shapiro, J. 2001. Mao's War Against Nature: Politics and the Environment in Revolutionary China. Cambridge: Cambridge University Press.
Scott, Alan. 2019. "City-regions Reconsidered." *Environment and Planning A* 51 (3): 554–580. doi:10.1177/0308518X19831591.
Sharma, Mihir S. 2015. *Restart: The Last Chance for the Indian Economy.* Gurgaon, Haryana: Random House India.
Smith, Neil. 2004. *American Empire: Roosevelt's Geography and the Prelude to Globalization.* Berkeley, CA: University of California Press.
Soja, Edward. W. 1989. *Postmodern Geographies: The Reassertion of Space in Critical Social Theory.* New York: Verso.
Tang, Wing Shing. 2014. "Governing by the State: A Study of the Literature on Governing Chinese Mega-cities." In *Branding Chinese Mega-Cities: Strategies, Practices and Challenges,* edited by P. O. Berg and E. Björner, 42–63. Cheltenham: Edward Elgar.
Tang, Wing Shing. 2017. "Beyond Gentrification: Hegemonic Redevelopment in Hong Kong." *International Journal of Urban and Regional Research* 41 (3): 487–499. doi:10.1111/1468-2427.12496.
The Geographical Society of China. 2018. "Introduction to the Society." Accessed 8 August 2019. http://www.gsc.org.cn/channel.aspx?id=5
The Royal Geographical Society. n.d. "The Society: History of the Society." Accessed 1 August 2019. http://www.rgs.org/AboutUs/History.htm
Tomba, Luigi. 2017. "Gentrifying China's Urbanization? Why Culture and Capital Aren't Enough." *International Journal of Urban and Regional Research* 41 (3): 508–517. doi:10.1111/1468-2427.12494.
Walder, Andrew. 2004. "The Transformation of Contemporary China Studies." In *The Politics of Knowledge: Area Studies and the Disciplines,* edited by D. L. Szanton, 314–340. Berkeley: University of California Press.
Wang, Qing Hui, Qi Wang, and Nian Cai Liu. 2011. "Building World-Class Universities in China: Shanghai Jiao Tong University." In *The Road to Academic Excellence: The Making of World-Class Research Universities,* edited by P. J. Altbach and J. Salmi, 33–51. Washington, D.C.: World Bank.
White, James W. 1998. "Old Wine, Cracked Bottle? Tokyo, Paris, and the Global City Hypothesis." *Urban Affairs Review* 33 (4): 451–477. doi:10.1177/107808749803300401.
White, James W. 2011. *Mirrors of Memory: Culture, Politics, and Time in Paris and Tokyo.* Charlottesville, VA: University of Virginia Press.
Wimmer, Andreas, and Nina Glick Schiller. 2002. "Methodological Nationalism and Beyond: Nation–State Building, Migration and the Social Sciences." *Global Networks* 2 (4): 301–334. doi:10.1111/glob.2002.2.issue-4.
Woldenberg, M. J. and Berry, B. J.. 1967. "Rivers and Central Places: Analogous Systems." *Journal of Regional Science* 7 (2): 129–139. doi:10.1111/j.1467-9787.1967.tb01429.x
Woldenberg, Michael. J. 1979. "A Periodic Table of Spatial Hierarchies." In *Philosophy in Geography,* edited by G. Gale and G. Olsson, 429–456. Dortrecht: Springer Netherlands

Wu, Fulong. 1997. "Urban Restructuring in China's Emerging Market Economy: Towards a Framework for Analysis." *International Journal of Urban and Regional Research* 21 (4): 640–663. doi:10.1111/ijur.1997.21.issue-4.

Wu, Fulong. 2016. "China's Emergent City-Region Governance: A New Form of State Spatial Selectivity through State-orchestrated Rescaling." *International Journal of Urban and Regional Research* 40 (6): 1134–1151. doi:10.1111/ijur.v40.6.

Wu, Zhong. 2011. "New Leftists Salute Wall Street Brethren" (*Asia Times Online*, 13 October, 2011). Accessed 1 August 2016. http://www.atimes.com/atimes/China/MJ13Ad01.html

Xin, Wang, and Qihong Liu. 2001. "Sanda Moshi, Hequhecong (What Lies Ahead for the Three Major Models)." *Zhongguo Jingji Kuaixun (China Economic News)* 2001 (29): 13–17.

Xinhua. 2015. "Chinese Communist Party Publishes 'Self-decoding' Series for Western Audience," May 29. Accessed 8 August 2019. http://en.people.cn/n/2015/0529/c90777-8899515.html

Xinhua. 2018. "China Focus: Xi Calls for Developing China into a World Science and Technology Leader," May 29, Accessed 8 August 2019. http://www.xinhuanet.com/english/2018-05/29/c_137213175.htm

Yang, Mayfair, ed. 1999. *Spaces of Their Own: Women's Public Sphere in Transnational China*. Minneapolis, MN: University of Minnesota Press.

Zhu, Tao. 2011. "Building Big, with No Regret." *AA Files* 63: 104–110.

Uncorking the neoliberal bottle: neoliberal critique and urban change in China

Will Buckingham

ABSTRACT
Neoliberalism's theoretical ascendancy within urban geography coincided with the rapid growth of scholarly attention to Chinese cities. Therefore, it is unsurprising that neoliberal causality has been a widely used tool for interpreting China's spatial transformation. This paper critically reviews some of the most prominent debates on neoliberalism in the Chinese context. China's Leninist political hierarchy and Dual Structure, crucial institutions for the management and regulation of society and economy under Mao, are now reduced to the quirks of "actually existing neoliberalism." Neoliberal critique applied to China, however, fails to adequately explain China's spatial development because it assigns causality for social and economic inequality to globalized processes of capital accumulation while ignoring the continued importance of Maoist institutions in China's present-day political economy. Uncritical acceptance of neoliberalism's explanatory power for spatial change has led to flawed and inaccurate portrayals of the development and future trajectories of Chinese cities, and misrepresents the sources of social injustice in Chinese society.

Introduction

Under the assumption that capitalism's ascendancy in China is inevitable, if not complete, many scholars employ a structuralist framework of capitalism to explain its spatial change. According to this logic, China's present-day spatial organization is the outcome of a globalized process of capital accumulation where the state engages primarily as the organizer of "urban entrepreneurialism" and place-making to attract capital. This approach is appealing because it links China research with an established tradition of critical scholarship, loosening the bonds of "Chinese exceptionalism" that have long plagued Area Studies. In short, neoliberalism gives urbanization a familiar and recognizable shape in a wide variety of local contexts.

Yet, it also limits how complex social relations that give shape to urban spaces are imagined. For Gibson-Graham (2006, 4) the identification of neoliberalism in diverse locales "affords the pleasure of recognition, of capture, of intellectually subduing that one last thing; it offers no relief or exit to a place beyond." Strict adherence to capitalist spatial logics leaves many features of China's urban landscapes unexplained. The uninterrupted rule of the Chinese Communist Party (CCP), no matter how decentralized and even fragmented its authority might be, continues to dominate China's political and economic order, preventing free market economics from achieving the ideological hegemony that is often presumed in the English-language literature. Moreover, the legacies of state socialism continue to dominate China's spatial political economy in the form of the dual structure that divides rural and urban China. Neoliberal critique is yet to adequately address the foundational role Maoist institutions continue to play in the regulation and organization of Chinese space. To understand the development of Chinese cities, we must explore the complex interrelations among agents nested in state hierarchies, among cities and individuals inside and outside China, and the strong and weak networks that constitute social life and economic activity.

Neoliberal critique and the geography of China's development

Following the calls of Brenner and Theodore (2002) and Peck and Tickell (2002) for greater attention to the intersection of neoliberal discourse and practices of urban governance, urban geographers have increasingly identified cities as the critical site of neoliberalism's implementation. Within China studies, the preponderance of neoliberal theorizing has coincided with increased attention to the role of cities in China's changing political economy (Tang and Chung 2002; Wu 2004; McGee et al. 2007; Wu 2007; Ma 2005). Given these overlapping agendas within China and urban geography, locating the intersection of neoliberalism and city development is an inevitable target of China scholarship.

However, there are some inherent contradictions between a theory that gives primacy to the market as the ultimate arbiter of sociopolitical conflict with private property rights as an ultimate social good, and China, a country where the market arguably remains subjugated to the priorities of the central and local state and all land is legally owned either by the state or rural collectives (while remaining effectively under the control of the CCP). Yet neoliberalism is frequently employed by China geographers with little critical debate over its basic assumptions. The nuances of its diffusion, implementation, and evolution are debated, but its essential character is rarely called into question *within* geography. While there are a handful of exceptions (Gibson-Graham 2006; Barnett 2010), the tendency is toward two extreme poles where neoliberalism is either ignored altogether (as found with a good deal of quantitative geography) or assumed as *the* causal force for a wide range of urban affairs. Research calling neoliberalism's explanatory power into question or demanding greater precision in the substance of neoliberal critique takes place primarily in the pages of sociology and anthropology journals.

China and neoliberal hegemony

Framing Chinese development in the reform era under the umbrella of neoliberalism rests on the premise that Chinese policy has been ideologically swayed by the insidious spread of neoliberal ideology. At the heart of critical anti-neoliberal theory's ascendancy in geography is David Harvey. Harvey has written extensively over the past two decades on the ideological hegemony of neoliberalism throughout the globe, with a critical eye frequently turned toward China (Harvey 2003, 2005). Neoliberalism, for Harvey is

> ... a theory of political economic practices that proposes that human well-being can best be advanced by liberating individual entrepreneurial freedoms and skills within an institutional framework characterized by strong private property rights, free markets, and free trade. The role of the state is to create and preserve an institutional framework appropriate to such practices ... if markets do not exist then they must be created, by state action if necessary. *But beyond these tasks the state should not venture.* (2005, 2)

The state under neoliberalism must limit its market interventions due to the supposed superiority of market signals and the state's susceptibility to powerful interest groups aiming to distort the market. Due largely to political shifts under Thatcher in Great Britain and Reagan in the United States, and the opening and reform of the Chinese economy begun in 1979, Harvey (2005) argues that neoliberalism has achieved hegemonic status in shaping political institutions and policy decisions. It is first and foremost, a *global* ideology determining political discourse from the United States to Sweden, from Brazil to Malaysia, becoming "incorporated into the common-sense way many of us interpret, live in, and understand the world" (Harvey 2005, 3).

Anti-neoliberal critics of Chinese development, including Harvey, rely on several characteristics of the reform period that restructured scale-relations to conclude China is, if not a fully neoliberal country, at least headed in that direction (Walker and Buck 2007). Of particular interest are the processes of fiscal decentralization, the "privatization" of public goods as a means of primitive accumulation, and urban entrepreneurialism. Although, these processes bear only superficial commonality with similar trends in Western and other developing countries, the arguments follow a very simple, if tautological thread: neoliberalism over the past 30 years, in the words of Harvey (2005), has "swept across the world like a vast tidal wave of institutional reform and discursive adjustment." China *has* undergone sweeping institutional reform and discursive adjustment over the past 30 years. Therefore, China's institutional reform is neoliberal by nature.

Harvey's analysis, however, leaves quite a few dots unconnected. Following Michael Peter Smith's general critique of Harvey (Smith 2001), Harvey's focus on the global scale creates a lack of specificity that overlooks the local political and social relations that mediate class relationships. In *A Brief History of Neoliberalism*, we see an inattention to the structure and agency of the state and other social actors that conflates the mere existence of capitalist interests with the full realization of neoliberal capitalism (Smith 2001). Corresponding with the emergence of corporate power in China, a disenfranchised and exploited labor class is also visible (Chan, Ngai, and Chan 2010; Chan, Ngai, and Selden 2013; Gong 2014). However, corporate power and labor dynamics are imbued with profound social, cultural, and political dynamics that are not captured by the linear diffusion of neoliberalism across space.

Anthropological skepticism toward neoliberal hegemony

Donald Nonini (2008) argues that the only evidence actually offered by Harvey for the case of neoliberalism in China is that its market reforms coincided with a neoliberal turn in the United States and Great Britain. While there are most certainly neoliberal *influences* at play in China, particularly among certain elements of the ruling class and economic elites, neoliberals represent only one of many voices creating domestic economic policy. Even among economic elites benefitting from privatization and liberalization within the Chinese economy, it is arguable whether or not they retain enough independence from the Chinese state to truly be considered part of a transnational business class that pays no heed to the reference of international borders suggested by Castells (2011) and Sparke (2005), among others (Dickson 2003; Chen and Dickson 2010). Hoffman (2010), for example, shows how many professional Chinese use international connections, especially education and training in Western universities to supplement, not replace, their identity as Chinese citizens. Many of the wealthiest people in China are either directly or indirectly supported by the party-state and continue to rely on connections to the state to expand business empires (Dickson 2007).

Nonini (2008) is critical of what he sees as a general conflation of *neoliberalism* with *capitalism*. The difference for Nonini is that neoliberalism is inherently about transforming governing practices within existing capitalist states, specifically "the relationship between capitalist markets, the individual, and the state" (149), whereas what is underway in China today is the creation of economic, technical, legal, and political arrangements to "invent capitalism." We might also extend Nonini's critique to include a conflation of capitalism with markets. One of the first events to herald changing times under Deng Xiaoping was the reemergence of spontaneous farmer's markets in Chinese villages, towns, and even cities. The availability of alternative and more diverse sources of food and other goods coming from the countryside in urban areas weakened the state's control over the allocation of goods and, especially price controls, but such markets operated independently and differently from state-regulated capitalism that would develop under Jiang Zemin in the 1990s. The emergence of flexible market practices in some areas of China, therefore, does not *ipso facto* translate into neoliberal restructuring simply because Reaganomics and Thatcherism were simultaneously rolling out in the West. That China's current system is conducive to accumulation does not mean that it is defined by it – capitalism "does not dictate its own conditions of existence" (Nonini 2008, 150) – it can and does operate in a variety of contexts, but its causal impact on states, cities, and cultures is not predetermined. Capitalism need not be fetishized as immune to the contextualization that any other social phenomenon is routinely subjected to in critical scholarship. Capitalism is acted on by local agents, subjected to manipulation by diverse interests with motives and desires that may or may not be strategically aligned with capitalist goals. It is out of the interrelations among a multitude of actors, institutions, and ideas, not the simple variations of global capitalism that the Chinese city emerges.

Accepting the hegemonic status of neoliberal discourses in China as a matter of course leads to misunderstanding the basic function of China's economy, faulty interpretations of reform outcomes, and misdiagnoses of the primary challenges facing continued social and economic transformation. For example, on the "proletarianization" of China, Harvey (2005, 144) states, "In so far as neoliberalism requires a large, easily exploited, and relatively powerless labor force, then China certainly qualifies as a neoliberal economy." This is spurious reasoning – neoliberalism "requires" a large exploitable labor force, China has a large exploitable labor force, therefore China is neoliberal. But how is this labor force kept exploitable? What techniques of political power are employed to create and sustain this class of workers? How do workers accommodate or resist their exploitation? For Harvey, it is the neoliberal "tsunami," an external force imposing its will on the Chinese labor force. Yet, China's labor force is kept powerless and exploitable not by the ravages of a neoliberal order, but by the resilience of the *hukou* system, an institutional foundation of China's decidedly non-market economy under Mao that maintains an unequal balance between the registered population of cities and rural areas. What is crucial to recognize is that the *hukou* system sustains a low-wage labor force for China's factories, but its greater function is in reducing state costs of urbanization. Resistance to *hukou* reform stems primarily from local governments fearing the social and fiscal costs of unregulated migration. Thus, it is as much a system of state regulation as it is of protecting capital interests.

One of the *hukou* system's key functions in the 1960s and 1970s was to serve as an institutional basis for the redistribution of wealth from rural to urban areas. This was accomplished by linking access to state-provided goods, especially grain rations, to household registration. In the present environment, access to public housing has been a key service linked to the place of *hukou* registration that has maintained a division of labor in the present context as well as a means of determining eligibility for certain types of employment (see Wang 2005). Furthermore, Huang (2008) argues that the central government put a credit squeeze on the rural sector in the 1990s, hindering entrepreneurial prospects for peasants. The lack of available credit and other barriers to entry in rural entrepreneurship created a condition where the only reasonable means of obtaining non-agricultural income for rural residents was through migratory searches for wage labor. The intense competition for wage labor that resulted combined with the institutional prejudices inherent in the *hukou* system to severely suppress wages for rural–urban migrants (Meng and Zhang 2001; Huang 2008, 122, 123). The unfair and unjust environment for migrant labor has given China a tremendous advantage for its participation in the global market (it is the primary reason for the so-called "China price"; see Harney 2008), but it is an outcome of state socialist policies seeking to *control* the market, protect urban residents, control population, and reduce state costs and obligations for urban workers (Chan 2009a, 2009b; Young 2013).

Neoliberal theory purports that local societies and political structures are so inherently weak that *any* engagement with capitalism inevitably leads to their assimilation (Smith 2002). Yet, China has taken advantage of globally predominant neoliberal trade policies without adopting a neoliberal stance for itself. Rural–urban inequality in China does not derive from privatization or the supremacy of free market ideology – quite the opposite. Huang (2008) points out that the 1980s was a period of wide-scale *marketization* of the countryside, and the outcomes lifted hundreds of millions of peasants out of poverty as they engaged in productive entrepreneurialism, creating new wealth and opportunities for improved livelihoods without significantly undermining the collective culture of the village. The shocking inequality that for many defines China's development model today did not emerge until the state reasserted its power to divert resources from the countryside into the city. Instead of a *tabula rasa* for the institution of free markets, the state had at its disposal a legacy of interventions in practices of land ownership, distribution, and access to public goods built over the preceding decades of state socialism. Especially after 1992, local governments proved adept at utilizing the peculiar spatial governing practices and policies of the Chinese state that directed the fruits of reform and development away from rural households toward municipal governments and quasi-private developers (Tang and Chung 2002).

However, for Harvey and others, a country need not be committed to capitalism at all in order to be in "the embrace of neoliberalism." The history of neoliberalism for Harvey (2007, 28) is an exploration of either "a Utopian project providing a theoretical template for the reorganization of international capitalism or as a political scheme aimed at reestablishing the conditions for capital accumulation and the restoration of class power." He goes on to claim that "the principles of neoliberalism are quickly abandoned whenever they conflict with this class project." Neoliberalism, then, when it is not a system of economic organization, is a convenient ideological tool with which a wealthy elite can easily subjugate the majority with a rhetoric of individual liberty and freedom. Harvey's sleight of hand here is a neat rhetorical trick. If a country is advocating and implementing policies in accordance with neoliberal doctrine, then we can reasonably assume it is neoliberal. If it is advocating and implementing policies that diverge from neoliberal doctrine, then it is also neoliberal, so long as its ultimate aim is to divert resources into the hands of a privileged class. Neoliberalism, in this vein, is an elite strategy for the control of wealth; empty rhetoric to justify "class restoration."

The problem is, different countries have different class histories, and the basis for conflict over class claims on national resources frequently rests on legacies of the Maoist social order. Indeed, class conflicts that have emerged in recent years are often waged between social groups and state representatives. This can be witnessed in struggles, for example, between state workers and reformers of state-owned enterprises, villagers and village cadres, and migrant workers and police. These contradictions cannot be reduced to oversimplified relations among capital, bourgeoisie, and the proletariat, but one must be cognizant of how these factors are interwoven with the social and political identities constructed under Maoism. But for Harvey (2005), local difference does not alter the core truth of neoliberalism, it may only change its superficial form, becoming, in his words, "neoliberalism with Chinese characteristics." As much as recent Marxist geographers have deplored the aspatial nature of neoliberal advocates like Thomas Friedman, the concept of space and local difference in Harvey's vision of neoliberalism reduces local difference to dependent variables of the particular form neoliberalism will take in different national contexts, or economic reductionism on a global scale. Pickvance (2011, 56) succinctly unsettles Harvey's interpretation of Chinese neoliberalism by pointing out that "to call it neoliberal with Chinese characteristics is to fail to see that the Chinese characteristics are in contradiction with neoliberalism."

Actually existing neoliberalism with Chinese characteristics

"Neoliberalism with Chinese characteristics" is used by Harvey to acknowledge the adaptation of neoliberalism to China's specific political economy. I find the phrase, along with its many variants, disingenuous; a means of exonerating the author from explaining local imperfections of the neoliberal model, subordinating local agency to the hegemony of neoliberal urbanism rather than exploring the specific processes through which it is operated. Such a framework is the descendant of Brenner and Theodore's (2002) call to study "actually existing neoliberalism" that aims to give greater attention to the specific processes through which neoliberalism works in divergent societies, especially at the urban scale. This had a profound impact on urban scholars of China who could now readily link their empirical work to a theoretical package with universal claims, overcoming the criticism of "Chinese exceptionalism" in China scholarship. It also gave influential geographers the authority to describe China's reform and development as "decidedly" neoliberal despite apparent contradictions (Harvey 2005; Ma 2005) and gave China's urban scholars a renewed focus on processes of primitive accumulation and urban entrepreneurialism (McGee et al. 2007; Lin 2009).

It is my view that "actually existing neoliberalism" is among the most detrimental concepts to emerge in urban geography in recent decades. In practice, it serves as little more than pretense to explain away local differentiation as part of a generalized patchwork of capitalist practices. As Brenner, Madden, and Wachsmuth (2011) criticize assemblage for decontextualizing cities from global capitalism, "actually existing neoliberalism" decontextualizes cities from their individual histories and relations, which are subsumed as variants of capitalism and capitalist structures. He and Wu (2009, 283) argue that "actually existing neoliberalism" is a useful framework for interpreting the production of urban space, as cities "have become increasingly important geographical targets and institutional laboratories for various neoliberal experiments, e.g. place-making, urban development corporations, public-private partnerships, new forms of local boosterism, and property-led redevelopment." The last of these, property-led redevelopment, is the focus of their 2009 paper, approached implicitly as a process of primitive accumulation whereby ownership of collective land is wrested away by private development companies. He and Wu (2009) consider three elements of China's urban development as decidedly neoliberal: the shifting of decision-making to local state authorities (decentralization), commodification of urban land, and the privatization of housing provision. While the former is an explicitly political move, it has facilitated accumulation through the commodification of land and privatization of housing. Shifting local decision-making and responsibility for revenue generation encouraged local political units to become economic decision-makers and fostered urban entrepreneurialism. Urban development, "which used to be the obligation of the state," they argue, "has now been transferred to the pre-mature market" (290). The resulting transformation of land and housing corresponds directly with central features of traditional accumulation: the commodification and privatization of land, the forceful expulsion of peasant populations, and the conversion of various forms of property rights into exclusive private property rights (Harvey 2001).

Inserting "neoliberal urbanism" as explanation for urban development does not do justice to the complexities of land acquisition and development that takes place within a decidedly unique, local context. Here, we see the fundamental flaw of what Tang (2014) calls "random conceptual indigenization." That is,

> searching for those Western concepts that seem to depict approximate patterns, and applying them to urban China. A Western concept is chosen merely on the basis that it exhibits more or less similar descriptive features to the concerned Chinese phenomenon. (47)

The resulting studies do not explain Chinese difference. Instead, their primary purpose is to explain those differences away:

> Not having undertaken a critical understanding of the Chinese context and processes, the China researcher conceives of urban space in China as if it were a spatial container that we fill with features identified by the Western concept *rather than perceiving it as a structure that the context and processes help produce.* (47, 48, emphasis added)

The presentation of urban development in China as led by capital is a purely subjective and, in this case, arbitrary position defining the transfer of urban real estate to private developers as central, but the transfer of property and wealth within the organization of the state is equally important.[1] Indeed, the claim that urban development is taking place under the *authority* of the market and not the state is premature. The development of commercial real estate that supposedly integrates Chinese territory with global circuits of capital is at times incidental to the opening of new spaces of state-sponsored and imagined urban modernity and the local state's imperative to generate revenue for social spending and other centrally mandated projects for local governance. Understood in this light, capital accumulation in China represents a transfer of wealth from individuals and collectives *to the state*. Yet, there is no contradiction between a theory that purports to shift state resources to private ownership and a state practice that manipulates its property rights regime to ensure *state* accumulation, because once capital "penetrates" a society, its dominance is ensured. In other words, "China's neoliberal urbanization optimizes market operations and maximizes the interest of the state-led regime of accumulation" (He and Wu 2009, 296). While China's urbanization may be highly problematic for conditions of equality, fairness, and social justice, the idea that it reflects optimal market operations is indefensible. The conditions under which land transfers take place guarantee the state's involvement and capture of revenues. How can a system be portrayed as neoliberal while simultaneously seeking as its primary aim to "maximize *state-led* accumulation," which is antithetical to the most primary tenets of neoliberal dogma? I believe there is a need to distinguish accumulation undertaken to advance state (or some other actor's) goals and accumulation undertaken purely for profit. Doing so would enable us to interrogate the assumption that the existence of accumulation is necessarily in support of capitalism or neoliberalization.

Without disputing any of the most basic claims about *how* urban development takes place in China, the utility of neoliberalism as an explanatory concept is lacking. He and Wu (2009) suggest that once we accept neoliberalism as *the* predominant paradigm of urban political economy, then we can readily link Western and Chinese logics of urbanism. They claim that "similarities can be found" between China and the West "in the most essential aspects of neoliberalization" and that this "essential aspect" is the "institutionalization of growth as the primary goal of the state" (299). They further point to similarities in local state tactics for neoliberalization, introducing "new elements such as capital subsidies, place-promotion, supply-side intervention, and local boosterism, into its central urban policy" (299). What we do find, however, is that China continues to rest on a mixture of "market logic and state authority logic." However, this state authority is only employed when it is necessary to balance neoliberalism's progress with societal contestation. While neoliberalism's progress in China is halting, its ascendancy is inevitable and irreversible.

Perhaps, it is the concept of a neoliberal essentialism that is most troubling about their arguments in that it effectively erases any significance of local variation for understanding how Chinese cities work. "Neoliberal theory," writes Daniel Goldstein (2012), "may be sketched on the backs of cocktail napkins from the Washington Hilton, but that is not the extent of it." Yet to deny neoliberalism its essentialist nature is not to trivialize capitalist influences on urban development. It is to investigate neoliberalism(s) as

> … not merely locally variegated instantiations of global ideas but fully lived realities in which people and states have their own theories, and elaborate their own discourses and critiques, about the worlds they inhabit and the ways in which these should be organised. "Actually existing neoliberalism(s)" are more than curious local manifestations of global norms, but sets of theories and practices about the world that are fundamentally the products of local history and experience … and impactful of lived daily reality. (2012, 305)

If there is not one core element of neoliberalism that can characterize a neoliberal city, then it is more fruitful to explore the diverse strands of intellectual discourse that embrace, reject, and adapt a wide set of theoretical, cultural, and moral beliefs into the production of urban spaces.

Spatial fixes: land, accumulation, and property rights

According to Harvey (2005), the role of the state under neoliberalism is to "create and preserve an institutional framework" of free markets for the exchange of private property. Asserting private property rights over public goods, which then monetizes collective property and integrates it with "global flows of capital" is the very definition of primitive accumulation (Harvey 2001; Glassman 2006). In China, the expropriation of rural land for urban construction is central to the current accumulation regime. George Lin, one of the most prominent geographers writing on China's urbanization and land development today, is strongly influenced by Harvey's work on "accumulation by dispossession"(McGee et al. 2007; Lin 2009). His work is most often empirical and is grounded in the experiences of the Pearl River Delta (Lin 1997, 2001, 2004). Lin's well-received 2009 book, *Developing China: Land, Politics, and Social Conditions,* is rich in empirical data, providing an ambitious overview of land development over the past two decades. Central to debates about the nature of Chinese capitalism, and a central focus of *Developing China*, is the definition and development of property rights.

China's property rights, especially with respect to land, are commonly understood as ambiguous. This ambiguity has been a focus of research on China's spatial changes over the past decade (Ho 2001; Whiting 2001; Ho and Lin 2003; Clarke, Whiting, and Murrell 2008). In particular, scholars have explored how agriculture, industry, and real estate have expanded despite the lack of full privatization advocated by classical economic theory. For students of political economy, the question becomes how China has managed to sustain capital accumulation without full privatization. Lin (2009) adroitly critiques the positioning of ambiguity within the property rights literature on China from two perspectives. First, in an approach clearly influenced by Douglas North, ambiguity is seen as the source of inefficiencies and social conflict because legal definitions of rural property rights are easily overwhelmed by the political power of the state (examples include Putterman

1993; Tian 2008). On the other hand, Peter Ho argues that ambiguity gives the state flexibility to adapt to markets, intervene when necessary, and maintain a large measure of control over development patterns (Ho 2001).

Harvey (2003) identifies land development as the primary "spatial fix" for capitalist over-accumulation, the means through which land is brought into the mainstream of capital accumulation. Lin (2009, 45) does not dispute the primacy of capital accumulation in spatial change, but he does argue that Harvey overestimates the role of global capital, suggesting that capital formation and sources of production and consumption are far more domestically driven. Still, land development for Lin is fundamentally driven by the "imperative of accumulation by dispossession." Here, the politics of land development revolve around coalitions of local state agents and private developers forming in competition with rural collectives for the capture of capital accumulation. I believe Lin misses the opportunity to explore an aspect of the relationship of capital, politics, culture, and space that are obscured whenever capital is given a dominant position in any localized "hybrid" of these layered social forces. Just because a phenomenon or local practice may be of use to global (or even local) capital accumulation does not mean that the phenomenon itself is caused or driven *by capital*.

Lin acknowledges that the central state has multiple goals for land development that *include* capital accumulation, but he does not address the motive for accumulation by the different actors. In Marxist land economics, accumulation occurs for its own sake, as a spatial fix for over-accumulation, but what is the motive for local governments and rural collectives to accumulate? What are they accumulating, and for what purpose? In standard political economy, accumulation is not seen as a strategy but as *the goal*. But is this always the case? Furthermore, Lin has created an overly simplistic division between urban governments and peasants, ignoring the interweaving of rural and urban society at the urban fringe and the tremendous transformations that have taken place within rural society. China's *de jure* rural population is too diverse for motives to be ascribed to a unified "peasant" class.

Developing China opens with a direct quotation from Marx on primitive accumulation, thrusting Marxist political economy, and by extension, Harvey, to the forefront of his analysis. Lin (2009) recognizes the hybrid character of China's land system and queries how land use and land systems are shaped by localized social, economic, and political conditions. However, due to the tautology of Marxist political economy, Lin cannot escape the limiting binaries of structure and agency and the historical inevitability of capitalist ascendancy. Thus, social conditions are determined in response to "historical change," economic conditions are confined to capitalist penetration, and political conditions are only understood in terms of central-local dynamics within the state. In each of these, a discrete and static "inside" is acted on by a dynamic external force; global capitalism acts on the state, which in turn acts on society to create the social conditions under study. Indeed, capitalism is regularly used as a synonym for history, the history that the Chinese state has rejoined following Mao's death. Before the 1978 watershed, China was a "rare" place where the "natural course" of the evolution of the economy and society were distorted and disrupted (2009, 72).

Land, as Lin recognizes, is subject to considerations beyond the scope of either classical economics or neo-Marxist political economy. Lin is right to assert that neoliberal frameworks for land development in China are too simplistic because in China, capitalist logics are mediated by cultural views of land as fundamental for social security, livelihood, and by the ideological prominence of land in the construction of CCP legitimacy. Land is not simply a commodity, he argues, but has multiple and conflicting meanings for individual farmers, village cadres, municipal governments, and the central state. Unfortunately, this assertion, as well as the substantive critique of "ambiguity" is soon forgotten as his analysis returns to a rational-choice model of resource contestation where social conditions are confined to responses to historical change, economic conditions are confined to capitalist penetration, and political conditions are only understood in terms of central-local dynamics within the state. In each of these, a discrete and static "inside" is acted on by a dynamic external force with global capitalism acting on the state, which in turn acts on society to create the conditions under study. Where he begins his work with a critical exploration of the assumptions about land and property in political economy, his own analysis becomes trapped by the limited agency allowed any force other than capital when accumulation by dispossession is taken as the driving force of history, and when central–local relations are reduced to a binary, scalar analysis.

A focus on material processes of land development obscures the intentions and outcomes of relevant actors who may engage or accept accumulation processes in order to achieve other desirable outcomes. It is thoroughly demonstrated that China's accumulation process encourages inequality, political oppression, and socioeconomic exclusion, but it is ontological fallacy to assume that the creation of capital surpluses shares a causal relationship with social injustice. More attention needs to be given to how ideas about social justice, equality, and social integration are constructed differently through localized relationships of space, power, and culture, and how the negotiation among those forces determines and legitimates the creation and distribution of capital surpluses (Gibson-Graham 2006). Instead of engaging these complexities, Lin too casually substitutes the agency of domestic capital for that of global capital.

Discussion and conclusion: uncorking the neoliberal bottle

In this paper, I have argued that cramming China's developmental experience into a neoliberal framework is fruitless for understanding that experience. It is the hallmark of Tang's (2014) "random conceptual indigenization." Moreover, the analyses identified above are quick to adopt the language of neoliberal critique, but they do not actually contribute to a more nuanced understanding of neoliberalism.

Instead, they simply force Chinese materialities into a recognizable, yet ultimately inadequate package of neoliberal urbanism.

As Dirlik (2012) suggests, however, it is not enough to simply declare that China is not neoliberal. To have theoretical relevance, urban geographies of China must be generative of new theories and new frameworks. While it is beyond the scope of this paper to do so here, I would like to close the paper by highlighting two points of emphasis that might begin to move urban geography away from "random indigenization and appropriation": history and the lived realities of urban life.

In the analysis of space, China geographers face a seemingly intractable problem of time. The linear and progressive view of history imbues nearly every commentary on China's developments. The existing focus on the diffusion of neoliberalism frames China as a generic spatial fix for crisis capitalism with market transition a catch-all for the particular forms of creative destruction underway. In the cases explored here, there is a burdensome assumption that socialism was China's past, and capitalism is China's future. Harvey's (2005) narrative says nothing about the pre-Deng era, as if China emerged from behind the bamboo curtain freed from three decades of Maoism, awaiting the welcoming arms of global capitalism. He and Wu (2009) describe an exploitative system of expropriation and dispossession as if these were inventions of neoliberalism rather than long-standing features that not only predate the reform era, but have roots in imperial practices of spatial governance. Most explicitly, George Lin (2009) characterizes China after 1978 as having "rejoined history." All of these examples emphasize an institutional context of neoliberalism that is utterly divorced from China's own institutional and historical context that must include the tremendous social, political, and economic transformations of the Maoist era for any understanding of what is actually happening in China today (Dirlik 2012).

When the influence of socialist institutions and the party-state are acknowledged, it is often characterized as path-dependence (i.e. "path-dependent capitalism" or "path-dependent neoliberalism"; see, for example, Wu 2010; Young 2013). While acknowledging that the past will influence the present and future, it is inherently misleading to describe history as path-dependent. It reaffirms the linear movement of countries through historical stages by allowing for a token amount of diversity in the final outcomes. But the path and history still have an end, and in China's case, this is largely seen as a quirky, yet ultimately recognizable variant of capitalism (Peck and Zhang 2013). The past is held as something that happened "before," and its lingering effects simply the result of institutional lag. That the contradictions of the present will dissipate as more fully capitalist institutions are established is the underlying assumption of market transition theory and has informed the research of prominent commentators on China (Nee 1989; Lardy 1998). It is a logic that is deployed to justify dispossession, expropriation, inequality, forced evictions, environmental destruction, and massive layoffs for the sake of a harmonious future. Under Mao, repression and violence against the people was similarly justified as necessary for building socialism.

Maoism and neoliberalism, then, both carry a promise that the violence and injustices of the moment will be erased once the next stage of history arrives. Yet the landscape of the present is shaped as much, if not more, by the past as it is an indeterminate future. For this reason, philosopher Manuel DeLanda (DeLanda, Protevi, and Thanem 2013) has argued for a "geological" approach to the city that not only recognizes but actively interrogates the historical "layers" of urban space. Here, social entities, including cities, are products of "specific historical processes, and whatever degree of identity they have, must be accounted for via the processes which created them and those that maintain them" (2013). The city (or for that matter, the region, nation, and so on) "emerges" as a whole from the multiplicity of its components (built environment, infrastructure, individuals, organizations and institutions, government), and must be continuously maintained by the interactions among the parts.

Theories about China's present must more consciously engage how it is built on the past. Lin's perspective erases three decades of Chinese history and leaves no room to consider how the transformation of China under Mao continues to shape the trajectories of its development, to say nothing of thousands of years of dynasticism. Holding the past and future as stark and contrasting moments in a teleological evolution of capitalism "shuts down the terrain of transition itself to critical inquiry" and risks overlooking the internal continuities, contestations, and mechanisms of social change (Martin 2007). Forcing Chinese cities and urbanization into the neoliberal bottle severely limits our ability to understand a relationship between past and present that is deeper and more complex than the simple institutionalization of capitalism.

There are existing examples of this kind of work that are sensitive to the continuities between so-called "pre" and "post" reform China, yet the degree to which they challenge the temporal bifurcation of China Studies is not widely recognized. Chan (2009b) has extensively detailed how the *hukou* system that developed in the first decade of the PRC has been foundational to China's ongoing economic transformations. Cartier's (2013, 2015) recent work has focused on how administrative divisions, many of which can be traced to imperial governing rationalities, shape the developmental trajectories of many of China's spatial units, showing how historical spatial practices produce the inter-city competition and urban entrepreneurialism typically attributed to neoliberal ascendancies. More than just an example of path-dependence, these works underscore a deep-seeded spatiality to Chinese governance that transcends our standard temporal markers and are necessary for any effort at understanding China's developmental trajectories.

Secondly, there is a notable dearth of investigations into how China's urban spaces are occupied, inhabited, and otherwise used in favor of cookie-cutter analyses that prioritize developers' intentions to "appeal to capital" (See Ong 2006, 2011 for a general critique). In short, the neoliberal intent of urban development is now all but taken for granted, but the unintended consequences of new urban spaces are rarely explored.[2]

Faced with a similar theoretical challenge, another group of regional urban scholars working in sub-Saharan Africa have launched a far more generative approach to urban studies. Despite obvious differences, African and Chinese cities share a long history of *in situ,* pre-modern urban spatial developments, a recent history of diverse experimentation with multiple modes of economic, spatial, and political governance, and a variety of contemporary practices that challenge the applicability of a neoliberal urbanism framework. In the provocative words of Parnell and Pieterse (2015, 241), urban scholars under these conditions are faced with a choice: "Either Africa must be ignored or the theory, method and data of urban studies must change." This caution has not been heeded in the China context. Rather than challenging the theoretical and methodological bases of Western urban studies, urban scholarship in China has settled for the neat packaging of neoliberalism where details and divergences are less important than the repetition of shared tropes of urban entrepreneurialism and capitalist penetration.

This also necessitates more direct engagements with the ways in which socialism continues to be constructed in China's development (Dirlik 2012; Lim 2014). Descriptions of Chinese socialism understandably tend toward cynicism, a reflection of both disillusionment with Maoism and skepticism over the concept of "socialism with Chinese characteristics." Yet, Lin (2006) draws on the ideals of Chinese socialism that exist independently of party elite constructions, ideals of justice, commitment, and social interdependence that inform contemporary contestations of state policy. Dislocation and expropriation described by He and Wu (2009) cast neoliberalism as the villain. Yet, amid the countless protests against expropriation across the country, capitalism itself is rarely an explicit target. Residents of China's villages-in-the-city do, however, frequently cite historical sacrifices, Maoist valorizations of peasants, and their own personal contributions to building Chinese socialism while simultaneously drawing on "market value" calculations in their claims against state-led expropriation of village property (Buckingham 2014). For the most part, residents' demands are modest in that they do not challenge the state's authority to expropriate, instead calling for more favorable relocation allowances. This is hardly the equivalent of anti-capitalist struggle (Harvey might describe it as "militant localism"), but it does reflect a framework for justice, rights, and fairness based on historical relationships between individuals and the state that is not recognizably Western. These relations are clearly not defined singularly by capitalism; rather, capitalism is integrated into a larger understanding of social relations connecting individual, community, state, and market actors.

Urban studies, whether centered in Europe or Asia, must be understood in the ways in which the city is embedded in structures, networks, and processes larger than itself (Cinar and Bender 2007). However, this is not to *reduce* the city to those external relations, which is what happens when urban geographers become singularly focused on identifying the neoliberal. We must be aware of *competing* narratives of the city that relate not singly and consistently to global capital, but also to nationalist projects, projections of state power, and cultural symbols that do not follow a narrowly Westernized conception of urbanism that is and always shall be construed as neoliberal (Ong 2011). In addition, the agency of residents of China's villages-in-the-city, migrant workers, and dispossessed farmers and their role in assembling the Chinese city deserves greater attention. At the same time, diverse motives drive the actions of state agents, party members, and the intellectual elite in both enabling and resisting acts of primitive accumulation. Greater attention to the ways in which neoliberalism is contested in China and the alternative tropes, ideologies, and worldviews actors employ not just in overt opposition to neoliberal policies but in the daily contestations over urban development and urban imaginaries could serve to decenter neoliberalism from Chinese analysis and give a more complete picture of how urbanism takes shape in different societies.

Notes

1. Hsing (2006, 2010), for example, has documented the tensions between governments and what she calls "socialist land masters," agencies, party and military offices, universities, and state-owned enterprises with direct access to large quantities of urban land. Additionally, there are sharp conflicts over land disposition among the Bureau of Land Management, Urban Planning Bureaus, townships, and municipalities. The transformation of land and real estate development is not simply a story of linear privatization; that is, the transfer of public goods into private hands. Instead, Hsing identifies a complex mix of multiple state actors, private and foreign owners, and socialist land masters who cause urban growth to be organized around competition and coalitions within the state.
2. There are outstanding exceptions including Ong (2011) on reactions to Rem Koolhaas' CCTV headquarters building in Beijing.

Disclosure statement

No potential conflict of interest was reported by the author.

References

Barnett, Clive. 2010. "Publics and Markets: What's Wrong with Neoliberalism." In *The Sage Handbook of Social Geographies*, edited by Susan J. Smith, Rachel Pain, Sallie A. Marston, and John Paul Jones III, 269–297. Los Angeles, CA: Sage.

Brenner, Neil, and Nik Theodore. 2002. "Cities and the Geographies of 'Actually Existing Neoliberalism.'" *Antipode* 34: 349–379.

Brenner, Neil, David J. Madden, and David Wachsmuth. 2011. "Assemblage Urbanism and the Challenges of Critical Urban Theory." *City* 15: 225–240.

Buckingham, Will. 2014. "Assembling the Chinese City: Production of Place and the Articulation of New Urban Spaces in Wuhan, China." PhD diss., University of Washington.

Cartier, Carolyn. 2013. "What's Territorial about China? From Geopolitical Narratives to the 'Administrative Area Economy.'" *Eurasian Geography and Economics* 54: 57–77.

Cartier, Carolyn. 2015. "Territorial Urbanization and the Party-state in China." *Territory, Politics, Governance* 3: 294–320.

Castells, Manuel. 2011. *The Rise of the Network Society: The Information Age: Economy, Society, and Culture*, 1 vol. Hoboken, NJ: Wiley.

Chan, Kam Wing. 2009a. "Population, Migration, and the Lewis Turning Point in China." In *The China Population and Labor Yearbook, Volume 1*, edited by Cai Fang and Du Yang, xix–xxxviii. Boston: Brill.

Chan, Kam Wing. 2009b. "The Chinese *Hukou* System at 50." *Eurasian Geography and Economics.* 50: 197–221.

Chan, Chris King-Chi, Pun Ngai, and Jenny Chan. 2010. "The Role of the State, Labor Policy and Migrant Workers' Struggles in Globalized China." In *Globalization and Labor in China and India*, edited by Paul Bowles and John Harriss, 45–63. London: Palgrave Macmillan UK.

Chan, Jenny, Pun Ngai, and Mark Selden. 2013. "The Politics of Global Production: Apple, Foxconn and China's New Working Class." *New Technology, Work and Employment* 28: 100–115.

Chen, Jie, and Bruce J. Dickson. 2010. *Allies of the State: China's Private Entrepreneurs and Democratic Change*. Cambridge: Harvard University Press.

Cinar, Alev, and Thomas Bender, eds. 2007. *Urban Imaginaries: Locating the Modern City*. Minneapolis: University of Minnesota Press.

Clarke, Donald, Susan Whiting, and Peter Murrell. 2008. "The Role of Law in China's Economic Development." In *China's Great Economic Transformation*, edited by T. Rawski and L. Brandt, 375–428. Cambridge: Cambridge University Press.

DeLanda, Manuel, John Protevi, and Torkild Thanem. 2013. "Deleuzian Interrogations: A Conversation with Manuel DeLanda and John Protevi." *Tamara Journal for Critical Organization Inquiry* 3: 1–36.

Dickson, Bruce J. 2003. *Red Capitalists in China: The Party, Private Entrepreneurs, and Prospects for Political Change*. Cambridge: Cambridge University Press.

Dickson, Bruce J. 2007. "Integrating Wealth and Power in China: The Communist Party's Embrace of the Private Sector." *The China Quarterly* 192: 827–854.

Dirlik, Arif. 2012. "The Idea of a 'Chinese Model': A Critical Discussion." *China Information* 26: 277–302.

Gibson-Graham, J. K. 2006. *A Postcapitalist Politics*. Minneapolis: University of Minnesota Press.

Glassman, Jim. 2006. "Primitive Accumulation, Accumulation by Dispossession, Accumulation by 'Extra-Economic' Means." *Progress in Human Geography* 30: 608–625.

Goldstein, Daniel M. 2012. "Decolonialising 'Actually Existing Neoliberalism'." *Social Anthropology* 20: 304–309.

Gong, Yue. 2014. "Manufacturing Towns in China: Governance, Space, and Conveyance of Rural Migrants to the Assembly Line." Diss., University of Washington. https://digital.lib.washington.edu/researchworks/handle/1773/25056.

Harney, Alexandra. 2008. *The China Price: The True Cost of Chinese Competitive Advantage*. New York: Penguin.

Harvey, David. 2001. *Spaces of Capital: Towards a Critical Geography*. London: Routledge.

Harvey, David. 2003. *The New Imperialism*. Oxford: Oxford University Press.

Harvey, David. 2005. *A Brief History of Neoliberalism*. Oxford: Oxford University Press.

Harvey, David. 2007. "Neoliberalism as Creative Destruction." *Annals of the American Academy of Political and Social Science*. 610: 22–44.

He, Shenjing, and Fulong Wu. 2009. "China's Emerging Neoliberal Urbanism: Perspectives from Urban Redevelopment." *Antipode* 41: 282–304.

Ho, Peter. 2001. "Who Owns China's Land? Policies, Property Rights, and Deliberate Institutional Ambiguity." *The China Quarterly* 166: 394–421.

Ho, Peter, and George Lin. 2003. "Emerging Land Markets in Rural and Urban China: Policies and Practices." *China Quarterly* 175: 681–707.

Hoffman, Lisa M. 2010. *Patriotic Professionalism in Urban China: Fostering Talent*. Philadelphia, PA: Temple University Press.

Hsing, You-tien. 2006. "Land and Territorial Politics in Urban China." *The China Quarterly* 187: 575–591.

Hsing, You-tien. 2010. *The Great Urban Transformation: Politics of Land and Property in China*. Oxford: Oxford University Press.

Huang, Yasheng. 2008. *Capitalism with Chinese Characteristics*. Cambridge: Cambridge University Press.

Lardy, Nick. 1998. *China's Unfinished Economic Revolution*. Washington, DC: Brookings Institution Press.

Lim, Kean Fan. 2014. "Socialism with Chinese Characteristics? Uneven Development, Variegated Neoliberalization and the Dialectical Differentiation of State Spatiality." *Progress in Human Geography* 38: 221–247.

Lin, George. 1997. *Red Capitalism in South China: Growth and Development of the Pearl River Delta*. Vancouver: University of British Columbia Press.

Lin, George. 2001. "Evolving Spatial Form of Urban-rural Interaction in the Pearl River Delta, China." *The Professional Geographer* 53: 56–70.

Lin, George. 2004. "Toward a Post-Socialist City? Economic Tertiarization and Urban Reformation in the Guangzhou Metropolis, China." *Eurasian Geography and Economics* 45: 18–44.

Lin, Chun. 2006. *The Transformation of Chinese Socialism*. Durham, NC: Duke University Press.

Lin, George. 2009. *Developing China: Land. Politics and Social Conditions*. Routledge Contemporary China Series. London: Routledge.

Ma, Laurence J. C. 2005. "Urban Administrative Restructuring, Changing Scale Relations and Local Economic Development in China." *Political Geography* 24: 477–497.

Martin, Patricia. 2007. "Mexico's Neoliberal Transition: Authoritarian Shadows in an Era of Neoliberalism." In *Contesting Neoliberalism: Urban Frontiers*, edited by Hette Leitner, Jamie Peck, and Erik Sheppard, 51–70. New York: Guilford Press.

McGee, Terry, George CS Lin, Mark Wang, Andrew Marton, and Jiaping Wu. 2007. *China's Urban Space: Development under Market Socialism*. Routledge Studies on China in Transition. London: Taylor & Francis.

Meng, Xin, and Junsen Zhang. 2001. "The Two-tier Labor Market in Urban China: Occupational Segregation and Wage Differentials between Urban Residents and Rural Migrants in Shanghai." *Journal of Comparative Economics* 29: 485–504.

Nee, Victor. 1989. "A Theory of Market Transition: From Redistribution to Markets in State Socialism." *American Sociological Review.* 54: 663–681.

Nonini, Donald. 2008. "Is China Becoming Neoliberal?" *Critique of Anthropology* 28: 145–176.

Ong, Aihwa. 2006. *Neoliberalism as Exception: Mutations in Citizenship and Sovereignty*. Durham, NC: Duke University Press.

Ong, Aihwa. 2011. "Introduction: Worlding Cities, or the Art of Being Global." In *Worlding Cities: Asian Experiments and the Art of Being Global*, edited by A. Roy and A. Ong, 1–15. New York: Wiley.

Parnell, Susan, and Edgar Pieterse. 2015. "Translational Global Praxis: Rethinking Methods and Modes of African Urban Research." *International Journal of Urban and Regional Research* 40: 236–246.

Peck, Jamie, and Adam Tickell. 2002. "Neoliberalizing Space." *Antipode* 34: 380–404.

Peck, Jamie, and Jidong Zhang. 2013. "A Variety of Capitalism ? With Chinese Characteristics?" *Journal of Economic Geography* 13: 357–396.

Pickvance, Chris. 2011. "The Limits of Neoliberalism: Is the Concept of Neoliberalism Helpful in the Study of Urban Policy?" In *Neoliberal Urbanism and Its Contestations: Crossing Theoretical Boundaries*, edited by J. Kunkel and M. Mayer, 46–62. Basingstoke: Palgrave MacMillan.

Putterman, Louis. 1993. *Continuity and Change in China's Rural Development: Collective and Reform Eras in Perspective*. Oxford: Oxford University Press.

Smith, Michael P. 2001. *Transnational Urbanism: Locating Globalization*. Hoboken, NJ: Blackwell.

Smith, Neil. 2002. "New Globalism, New Urbanism: Gentrification as Global Urban Strategy." *Antipode* 34: 427–450.

Sparke, Matt. 2005. *In the Space of Theory: Postfoundational Geographies of the Nation-State*. Minneapolis: University of Minnesota Press.

Tang, Wing-shing. 2014. "Governing by the State: A Study of the Literature on Governing Chinese Mega-Cities." *Branding Chinese Mega-cities: Policies, Practices and Positioning*, edited by Per Olof Berg and Emma Bjorner, 42–63. Northampton, MA: Edward Elgar Publishing.

Tang, Wing-shing, and Him Chung. 2002. "Rural-urban Transition in China: Illegal Land Use and Construction." *Asia Pacific Viewpoint* 43: 43–62.

Tian, Li. 2008. "The Chengzhongcun Land Market in China: Boon or Bane? – A Perspective on Property Rights." *International Journal of Urban and Regional Research* 32: 282–304.

Walker, Richard, and Daniel Buck. 2007. "The Chinese Road." *New Left Review* 46: 39.

Wang, Feiling. 2005. *Organization through Division and Exclusion: China's Hukou System*. Stanford, CA: Stanford University Press.

Whiting, Susan H. 2001. *Power and Wealth in Rural China: The Political Economy of Institutional Change*. Cambridge Modern China Series. Cambridge: Cambridge University Press.

Wu, Fulong. 2004. "Transplanting Cityscapes: The Use of Imagined Globalization in Housing Commodification in Beijing." *Area* 36: 227–234.

Wu, Fulong. 2007. *China's Emerging Cities: The Making of New Urbanism*. London: Routledge.

Wu, Fulong. 2010. "How Neoliberal is China's Reform? The Origins of Change during Transition." *Eurasian Geography and Economics* 51: 619–631.

Young, J. 2013. *China's Hukou System: Markets, Migrants and Institutional Change*. London: Palgrave Macmillan UK.

China's urban ideology: new towns, creation cities, and contested landscapes of memory

Tim Oakes

ABSTRACT
This paper argues that "the urban" has emerged as the most significant ideological realm in contemporary China. In developing this argument, I suggest an alternative approach to how we theorize urbanization in China. Seeking to avoid the dichotomized analyses that often characterize scholarship on China's urbanization, the paper suggests reading "the urban" as an ideological device. Such a reading calls for an analytical distinction between *the city* as a technology of socialist party-state planning and government and *urbanization* as a messy social process over which the state struggles for control. It also calls for a recognition of the ways that ideology itself has shifted dramatically in China, from the Mao-era centrality and coherence of class struggle and its overriding goal of proletarianization to a much less coherent post-reform message of "stability". The paper begins with a brief discussion of ideology and Gramsci's notion of "common sense" in a Chinese register. It then considers the film *24 City*, directed by Jia Zhangke, as a template for understanding urban spaces as sites of conflict between the city as an ideological device and urbanization as a social process. New urban spaces are then explored in an effort to tease out their complex and contradictory ideological renderings. I conclude with an argument about the openness and contradictions of China's urban spaces and how an ideological analysis can resist the kind of theoretical closure that much work on urbanization in China seems to aim for.

The city as memory/the city as history

> Today, as a Chinese person, I feel that it's urgent to confront the past, which China doesn't encourage … . I want to break with the official way of explaining the revolution that the Communist Party has appropriated, to bring to the fore my point of view as an artist and as a movie director.

The director Jia Zhangke recently made this comment in Paris while discussing his documentary history of Shanghai, *I Wish I Knew* (海上传奇2010), with Isabelle

---

This article has been republished with minor changes. These changes do not impact the academic content of the article.

Regnier of *Le Monde* (Le Monde 2011). With its inclusion of stories from Taiwan and Hong Kong-based Shanghainese, many of whom fled the city in the 1940s, *I Wish I Knew* recovers memories of the city that cannot be found within the space of the city itself. Shanghai is mostly depicted in the film as a city of rubble and massive transformation, a landscape of memories under assault and paved over. This theme of memory and the city is also the subject of Jia's 2008 film *24 City* (二十四城记). Both films offer meditations on the ways memory is held in the urban built environment, and how war (in the case of *I Wish I Knew*) or urban redevelopment (in the case of *24 City*) not only entail forced resettlement for the city's most marginal residents, but the evacuation of their memories as well. *24 City* records the demolition of a former military factory in Chengdu, and its replacement by a high-end commercial real estate development, all told through the memories of the factory's (former) workers and their adult children. For Jia, the destruction of urban landscapes is not a question of heritage preservation, but of the mundane everyday landscape that contains within it the memories of people's everyday lives there (c.f. Boyer 1994). Pierre Nora (1998) writes of "lieux de memoire": sites, landscapes, or objects that hold collective memories. These become significant sites of contestation over history and memory because while they may be the basis for collective memories, they might also be appropriated by governments or other powerful actors to become sites not of memory but of *history*, that is, a particular telling of the past that becomes a *substitute* for memory. And this process of state-induced amnesia (Lim 2014) is what is at stake in many of Jia's films.

I want to begin this paper on the urban ideology in China with the unexceptional observation that the built environment plays a role in contests over memory and history. This observation is the crucial starting point, though, for an argument about the city as an ideological device in contemporary China. China's cities are unsettlingly ephemeral. They are landscapes of massive demolition, topographies of rubble, templates for unprecedented social transformation. A central trope in Jia's films is the sense of disorientation that accompanies everyday life in urban China and how it can prompt a retreat into fantasy. This is seen, for example, in the animated interludes that illustrate one of the central character's reveries in his film *The World* (世界2004). "Most individuals in China are living in an illusory world," Jia says. "Reality is a mess in China. It's total chaos" (Zhang 2010, 89). It's a situation ripe for the imposition of a coherent narrative – an official history, a fantasy – powerful enough to quell and channel the wandering ghosts of memory unleashed by all the destruction. Controlling the narrative has become one of the most important tasks for the state. And the urban landscape has become perhaps the most prominent medium through which that narrative is both articulated and contested. The urban landscape is central to Jia's cinematic project of recovering memory and individual imagination. "Remembering history is no longer the exclusive right of the government," he says. "As an ordinary intellectual, I firmly believe that our culture should be teeming with unofficial memories" (Wu 2011, 9).

If the city is a contested space between chaotic memories of individuals on the one hand and coherent state-sanctioned histories on the other, then it is an *ideologically* contested space. Indeed, as I will argue in this paper, "the urban" has emerged as the most significant ideological realm in contemporary China. For the Chinese state, both urban "newness" and "heritage" are meant to eclipse heterodox memories of past errors and extremes (think pitched street battles between rival Red Guard factions during the Cultural Revolution, or soldiers shooting into crowds of residents blocking their march to Tian'anmen Square in 1989). While the Communist Party's primary ideological message has shifted significantly (class struggle no longer serves as the guiding principle for correct political thought), it is no less focused on monopolizing the values upon which social transformation should be based. During the Mao era, Party-state ideology reached its apogee of spatial expression in the dual organizational institutions of the rural commune (Howard 1988) and the urban work unit (Bray 2005). Today, the ideological realm remains powerfully defined by this rural-urban dualism. But the scale has shifted, and the content rendered more abstract, such that "the urban" and "the rural" have emerged as core, if not ambiguous, ideological concepts. But it is "the urban" that constitutes the space through which the Party's underlying values are to guide political thought and practice. "The rural" remains a derivative category: the "non-urban" other.[1]

In suggesting that we pay more attention to cities as ideological spaces in China today, this paper hopes to advance a broader argument about how we theorize Chinese urbanization, something that geographers have debated considerably over the past decades (Buckingham 2017; Cartier 2009; Chan 1994, 2007; Gaubatz 1996; Hsing 2010; Lin 2011; Ma and Fulong 2005; Fulong 2009; Wu and Gaubats 2013). While this literature is nuanced and highly attentive to the empirical realities of Chinese urban development, there remains much uncertainty about how best to theorize urbanization in China today. A key question seems to be how to balance, on the one hand, a focus on global patterns of urbanization and how they are capable of overwhelming local histories and institutions (i.e. "neoliberal Chinese cities"), with, on the other, the resilience of the local and its ability to cross-breed those global patterns into unique practices (i.e. "neoliberal urbanization with Chinese characteristics"). This balancing tends to result in dichotomized approaches emphasizing categorized distinctions between, say, urban and rural, state and village, formal and informal, legal and illegal. For example, urbanization in the Pearl River Delta region can be viewed as the hybrid result of a state-village developmental relationship, but we too often dichotomize that relationship by viewing the village as a non-state "society" actor, thereby reinforcing an urban-rural dichotomy in institutional terms. This paper emerges from a sense of dissatisfaction with these approaches to urbanization as a political-economic contest between "global" and "local" social forces. How might we theorize urbanization in ways that avoid scalar and institutional dichotomies? Or in ways that keep our focus on the continuities of Party-state governance, as opposed to the disruptions of neoliberal capitalism? How do we understand Chinese urbanization as constituting more than the "Chinese characteristics" of planetary urbanization?

Reading "the urban" as an ideological tool and strategy offers a promising response to these questions. Urbanization in China is understandably associated with a broader set of social changes instigated by post-Mao market-oriented reforms. But often along with this come assumptions about the reforms as a sort-of "ground zero" for the emergence of a new, globalized and quasi-capitalist China, accompanied by the ideological vacating of the Party-state (FloraCruz 1994; Harvey 2005). Rather than finding in urbanization the outcomes of globalizing processes of neoliberalism, or of state policies focusing on development and modernization, urbanization might well be understood as part of an on-going ideological project of shoring up state power and authority in a rapidly changing society. This is not to deny, of course, the significance of neoliberal capitalism or of, say, local-state land-based revenue generation as key variables in any explanation of Chinese urbanization. My aim is not to argue against the validity of other approaches so much as advocate for an alternative perspective focused on ideology, at a time when we tend to assume the ideologically heady times of Mao are long gone and replaced by a more pragmatic focus on GDP growth. Such an alternative approach might, then, help shed some light on the changing nature of ideology in China. It might help us better understand urbanization in China less as the spectacular manifestation of China's emergence in the world economy – an emergence that has in various ways challenged the state's ability to "control the narrative" and maintain ideological hegemony – and more as part of a process of state readjustment and the reproduction of state authority. Cartier asks, in her article in this issue (2019), "why do paradigms adopted for research on China derive more often from the experience of the urban process under capitalism rather than its legacies and transformations from the history of the planned economy under socialism?" Like the other articles in this special issue, this paper offers one attempt to redress that imbalance in analysis.

Seeing urbanization as ideological also helps us understand ideology less as the relatively coherent state-directed narrative that it was during the Mao era and more as a multivocal discourse with multiple actors and participants. This does not make it less powerful, or less contested. But it perhaps makes ideological practice in China today less certain and decided. Jia Zhangke's films – *24 City* and *I Wish I Knew* – suggest that the urban landscape upon which today's Party-state ideology is manifest is open and, indeed, incoherent, unpredictable, unstable even. For Jia, it is not a foregone conclusion that the marginalized and their memories will be bulldozed into piles of rubble. This is because urbanization as a messy social process undermines and destabilizes "the urban" as an ideological device, something that I will discuss more below. But what is equally important about considering the city as an ideological device is that it recovers some analytical continuity in how we understand the urban condition in China. Much of that continuity derives from an appreciation of the enormous revolutionary transformation that Chinese society underwent *prior to* the enormous urban transformation that it is undergoing today. "Newness" has been an ideological trope all along; it is not something that began with the new cities, new areas, and special economic zones of the reform and post-reform eras.

In this paper, then, I explore the city as an ideological device in China today in an effort to suggest an alternative approach to understanding urbanization, one that highlights the analytical continuities of (pre- and post-reform) socialist state-making at a time when much of our analytical attention has been preoccupied with the political economies of neoliberalism and urban entrepreneurialism "with Chinese characteristics." At the same time, though, I will argue that ideology itself has shifted dramatically from the Mao-era centrality and coherence of class struggle (阶级斗争) and its overriding goal of proletarianization (symbolically and materially realized in the body of the modern factory worker) to a much less coherent post-reform message of "stability" (维稳). As a new space of luxury housing, shopping malls, pedestrian streets, "green" and "smart" infrastructures, and well-behaved people of high quality, then, the city has become the epitomized landscape of social stability, home to the stabilizing bodies of high-rise-living middle class consumers. This shift, as I will discuss below, provides the narrative thread for Jia's film *24 City*. The paper proceeds with a brief discussion of ideology and Gramsci's notion of "common sense" in a Chinese register before considering *24 City* as an ambivalent meditation on the urban ideology. The film's deliberate confusing of reality and fantasy, its pitting of history against memory, and its insistence that nothing in urban China is quite what it appears to be, offers a template of sorts, I suggest, for how we understand new urban spaces in China today. These spaces – new towns, special economic zones and new areas – are then explored in an effort to tease out "the urban" as a powerful ideological device, but one with many agents, authors, and participants, none of whom have much control over the narrative. I conclude with an argument about the openness and contradictions of China's urban spaces and how an ideological analysis, counterintuitively perhaps, can resist the kind of theoretical closure that much work on urbanization in China seems to aim for.

The city as common sense

The rapidity and scale of China's urbanization has inspired awe and anxiety around the world. Vaclav Smil (2014, 91) famously claimed that China used more cement between 2011 and 2013 than the U.S. used in the entire 20th Century, a statistic that Bill Gates found "staggering" (Swanson 2015) and David Harvey found indicative of global capital's intrinsic crisis of accumulation (Harvey 2016). In 2013 *The New York Times* noted that over the coming decade China would move a mind-boggling 250 million people into cities (Johnson 2013). Its report was accompanied by an animated simulation of a nighttime satellite's view of earth from space. Glowing cities passed underneath while a digital caption counted out the equivalent number of existing cities that would need to be added to China's urban population in order to reach the government's urbanization target. The animation implied visually that China's urbanization is of a scale that is completely unprecedented, that can only be understood as planetary.

While it may therefore be entirely appropriate to consider China's urbanization within the framework of "planetary urbanization" (Brenner 2014; Brenner & Schmid 2011), the *spectacle* of the urban explosion in China tends to overwhelm much of our analysis, implying China's *sudden* arrival on the world's urban stage. To push the implication further, urban China has come into academic and popular media focus in a way that often isolates it from China's own historical, social, and political contexts, as if the reform era unleashed entirely new forces and forged connections that have moved China into an unprecedented new era, an urban revolution. The "newness" of China's arrival on the world's urban stage is a message that shares ideological affinities with the Chinese state's own narrative of a new, civilized, urban China.

In China, urbanization has come to dominate what Gramsci (1971) termed the "common sense". We see this in state planning and policy discourses – where China's developmental future, its ideas of sustainability, equity and justice, and its vision of itself as a prosperous, modern, and civilized nation – are all dominated by a sense of resolution in "the urban." Urbanism has become an ideological device in China by which all manner of social issues are understood, evaluated, and acted upon. The urban has colonized the imagination; it dominates how people contemplate the future, and how they judge the past.

While the spatial manifestation of this "common sense" is "the city", it is more specifically the *new* city, a space of redemption in which the mistakes and inefficiencies of older cities are resolved, and where the backwardness and poverty of the rural are substituted by the future and a middle-class lifestyle.[2] While these spaces are imagined and represented as entirely new and innovative, they in fact recall earlier spatial forms such as the model village and the socialist demonstration site. As exemplary, exceptional, and experimental spaces, these new areas have socialist and revolutionary genealogies, where bounded zones were created in order to demonstrate the social transformations that the rest of the world would someday also experience (Bach 2019). They might even echo, in some ways, earlier *imperial* orderings of space, such as the ideal palace city and the courtyard home archetype. But that may be stretching the genealogical connection too far. What can be safely claimed, though, is the will to order that these spatial formations share in common. That and a highly visible claim to territorial governing authority whereby the organization and bounding of space conveys sovereignty over land and populations (Cartier 2015).

As with these earlier forms, however, and as Jia Zhangke's films remind us, the urban ideology in China belies a more chaotic process of urbanization on the ground. Gramsci argued that the common sense, while dominated by the cultural hegemony of the ruling class, was always fragmentary and incoherent, a chaotic aggregate of disparate conceptions that contained within it the seeds of disruption and counter-hegemonic practice (Gramsci 1971, 419). While urban ideology conveys a China where state-of-the-art new towns built tabula rasa will redeem the socialist wreckages of the past, China's new urban areas are in fact chaotic spaces of accretion, haphazard appropriation, and popular hacking by unplanned-for users. They remain, as Sampieri (2019) suggests, uncertain and undecidable spaces due to their incompleteness. The boundedness of these new areas offers a spatiality that implies sharp lines of distinction between urban & rural, formal & informal, planned & unplanned, and unique & generic. But in fact, the spaces of the new urban zone can blur these distinctions, or even render them non-existent. The blurred distinctions and unexpected co-minglings that characterize the practices of urban zone construction in China today can be attributed, in part, to the ambiguities of the medium – the urban landscape – through which dominant ideologies themselves are being conveyed. One might say that the very *materiality* of the city – its infrastructures, its buildings, the routes of its commuters, their workaday lives – and the *memories* embedded in this materiality, provide for an openness that makes the city a particularly messy ideological device.

For the urban theorist Henri Lefebvre, the city was both an historical phenomenon – the outcome of social processes of urbanization – and an ideological representation. This duality emerged from the idea that while the city was a material manifestation of historical urbanization processes, those processes had, by the post-war era, exceeded the spatial format of the city itself. Critical analysis, Lefebvre argued, should thus focus on urbanization as a social process, rather than on the city as a spatial form. The modern urbanist commitment to the city as a space of planning and regulation was ideological in that it represented a project of abstracting away from, or even erasing, the social contradictions of urbanization. Lefebvre recognized the "instrumentality of urbanization processes in the reproduction of capitalist relationships, but at the same time he also explored urban space as a place of transgression and alternative social projects" (Schmid, Stanek, and Moravanszky 2014, 3). In *The Urban Revolution* (2003, 57) Lefebvre wrote that "an image or representation of the city can perpetuate itself, survive its conditions . . . In other words, the 'real' sociological 'object' is an image and above all an ideology!" Or, as Wachsmuth (2014) put it, "The city is an *ideological representation* of urbanization processes rather than a *moment* in them" (356, emphasis in original).

As a purely ideological device, then, the city could only exist as an image, a plan, a rendering. This is because the social processes of urbanization – and the lived experiences they generate – always exceed the conceived and planned formats of the city. This, we might say, is what makes the city a particularly messy ideological device. As will be discussed in a later section, the closest an actual city ever came to pure ideology, for Lefebvre, was the "new town," a "machine for living in" (Lefebvre 1995, 119). This offers a clue, perhaps, to the popularity of the new town format in China, as the state seeks to maintain control over the unruly and contradictory social processes of urbanization and as it seeks to institutionalize that control in the space of the city. It is also worth considering that in China "new town" developments are often built as planned and conceived, usually very quickly, allowing their ideological aspects to survive beyond the planning stage and become materialized in ways seldom achieved, for example, in Lefebvre's post-war France.

Still, the city has never been a particularly reliable ideological platform in China. Schurmann (1968) made this clear in his study of China's urban communes during the Great Leap Forward. Ideology was, according to Schurmann, the necessary basis of the Party's fundamental reorganization of Chinese social order; it was meant to completely replace the traditional "ethos" that held Chinese society together. Rural and urban reorganization was institutionalized in the commune and work unit, respectively, and each of these was imbued with a Maoist practical ideology emphasizing the struggle over, and resolution of, class contradictions. The modern industrial worker represented the embodied form of that struggle and resolution. Thus, for Schurmann, the objective of the commune was the "transformation of the village into a factory, and of the peasant into a worker" (41). This was a transformation based on values that were entirely unprecedented and unfamiliar in the rural Chinese social order.

There was, however, a strong tradition of urbanism in pre-revolutionary China. "The Chinese," Schurmann claimed, "have a strong affinity to urban life" (365). The ideological imperative for the revolutionary state was to reorganize urban life around processes of production, rather than consumption. The basic idea underlying the experimental urban communes, and of urban social reorganization more generally, was to bring production to the worker, rather than the other way around. "Residential blocks, they thought, could be transformed into production centers" (367). This "unity of living and working" was however achieved only sporadically, among a certain group of well-developed and well-funded industrial work units (such as the one, for example, featured in Jia's film *24 City*). Such work units represented the ideological apogee of the Mao era. But generally, China's cities remained, in Schurmann's opinion anyway, "great residential slums" whose ways of everyday life resisted ideological spatial reformatting. Revealing his own bias or skepticism about the redeemable qualities of modern urban life, Schurmann concluded that the Party had "not been able to create effective forms of social organization to bring about a new sense of community to replace the anomie inherent in modern cities with their crowded tenements and mixed populations" (402). An obvious reason for this lack of effectiveness was the state's limited resources. It had to rely on the Chinese city as an already-existing spatial form within which to ideologically reshape urban lives.

That constraint has become much less significant today. The Chinese state has greater capacity today to institutionalize its governing ideologies "in the social fabric of the city and [monumentalize] its symbolic values in the urban built environment" (Cartier and Tomba 2014). This has given "the city" much greater purchase as an ideological device than was previously the case. This did not happen all at once, however. During the 1980s and 1990s, a township-and-village-enterprise driven process of rural urbanization, or "urbanization from below" (Ren 2013), emerged as an unintended outcome of early rural reform policies. This was exactly the kind of urbanization process theorized by Lefebvre as exceeding the extant urban form. It is perhaps not surprising, therefore, that the state ultimately sought to rein-in this sort of unplanned and chaotic development. Rural urbanization "from below" threatened to eclipse the state sector which still served as the center's primary source of revenue, as well as the state's central institution of social control. But "urbanization from below" was a threat mainly because it undermined the core institution of the urban ideology: the state's categorical monopoly on distinguishing rural from urban. As Cartier and Tomba (2014) argue:

> An ideology is only as useful to domination as it is embedded in institutions and the urban itself has long been institutionalised. The classification of what is urban and what is not remains the most important and significant of these institutions, ever since the 1950s and remains an active marker to define the relationship of citizens with their community, their territory, the state and the economy. Despite the increasingly blurry boundaries of the 'city' – with, for example, most of the industrialisation and much of the rural-urban migration localised outside of the boundaries of traditionally defined urban areas – the framing of territories according to this institutional distinction remains a monopoly of the Chinese state and fundamentally shapes the destiny of every Chinese.

The 1992 designation of Pudong as a state-level new area signaled the beginning of the state's commitment to wresting ideological control over the urban away from the countryside, beginning what Ren refers to as an era of "top-down" or "city-centered" urbanization "characterized by decisions to invest in major urban projects by bureaucratic elites from the central and municipal governments" (Ren 2013, 29). While the state has maintained control over the rural-urban distinction through the *hukou* system of household registration, even as it reforms and makes this institution more flexible, it has also continued to insist on defining, precisely, what is "urban" and what is "rural" in an age where the *process* of urbanization has rendered that distinction increasingly meaningless in terms of everyday experience. Thus, the state invests an ideological distinction between rural and urban into institutions that otherwise serve the process of urbanization. Examples of this abound. The "New Socialist Countryside" campaign enabled the urban expropriation of rural land on a massive scale (Looney 2012) even as it promoted, ideologically, the drawing of a clear boundary between the rural and urban (by, for instance, promoting an aesthetic ideal of "the rural" via village beautification projects). Similarly, the "National New-Type Urbanization Plan" explicitly called for the preservation of rural culture. As articulated by one Chengdu official, "The new style rural urbanization is not meant to eradicate the countryside; rather, it is meant to make the city more like the city and the countryside more like the countryside" (Liu 2014, cited in Wilczak 2017, 111). Wilczak's (2017) study of rural-urban integration in Chengdu demonstrates how the incorporation of rural spaces within comprehensive urban planning jurisdictions was accompanied by a "pastoral aesthetic" that sought to, at least visually, maintain a rural-urban distinction even as "the rural" was disappearing as an autonomous institutional and social sphere.

While the urban has thus become ideologically central – a new "common sense" signifying everything that is modern, civilized, developed – urbanization has continued to produce spaces of everyday urban life that exceed, complicate, or contradict the urban ideology. The urban, as an ideological image, is constantly undermined by the process of urbanization. This, then, suggests an interpretive lens through which to explore the spaces of everyday life under urbanization in China, as conveyed in the cinematic vision of Jia Zhangke.

The city as performance art

In China, the baseline from which one's value as a state subject is gauged has shifted from industrialism to urbanism. This shift forms the narrative spine of *24 City*, a hybrid "docu-fiction" that chronicles the demolition of the Chengfa Group's Third Front combat aircraft engine factory in Chengdu (known simply as Factory 420) to make way for a sprawling commercial housing development called 24 City. Financed in part by the 24 City real estate development company, Jia's film is, ostensibly, a nostalgic homage to an era of collectivism, labor, and self-sacrifice to the nation. The retired and laid-off workers who tell their stories in long and patient soliloquies give voice to a socialist subjectivity in which the city was secondary to the work that went on within it. The Mao-era city of Chengdu was almost irrelevant to the self-contained world of Factory 420 which, due to the secretive and centrally-driven nature of its production, was functionally independent of the urban environment in which it happened to be located. Thirty thousand workers lived there, a "city within a city." Song Weidong (played by Chen Jianbin), raised in the factory and now a mid-level manager, recalls that "420 had almost no connection with Chengdu; we thought we didn't need the local people. No need. We had a great life. The only way we connected with local kids was through fights."

By the reform era, Factory 420 had shifted its production from military aircraft engines to refrigerators and washing machines. And by the late 1990s production had largely ceased altogether. In 2007 the land title was transferred from Chengfa to China Resources Land Ltd, in a staged ceremony held in the factory's auditorium with the remaining employees gathered to applaud this driving of the final nail into the coffin of socialist worker. This ceremony is how Jia's film begins. The factory will be replaced by a huge development meant to house 60,000 middle class residents, with gardens, parks, a shopping mall, and an 8-screen multiplex. While the name – 24 City – is said to come from an ancient poem about Chengdu, it evokes both a reversal of the old era (from '420' to '24') and a global-city coolness. "24 City" sounds like "24/7," a city that's always open, a city that never sleeps, nonstop entertainment and consumption, a city marching toward modernity at full speed.

Jia's film is full of the kinds of stark contrasts one would expect of such a profound transformation from socialist industrialism to global urbanism as the basis of one's value and identity. His workers – China's former heroes of production – are sad, almost pathetic souls; they are ghosts of their former selves, aimless and lost in a new world. They struggle to keep their livelihood, health, dignity, and mental faculties as the world passes them by. Most of them have nothing to do except dismantle their former workplace, piece by piece. Hou Lijun was laid off in 1994 and ended up selling flowers on the sidewalk, trying to keep clear of the "urban management" (城管) officials. She tells us: "if you have something to do, you age more slowly." Her mother came to Chengdu in 1958 when the factory was moved from Shenyang and never once returned to see her aging parents until 1972. The film is full of these stories of sacrifice. Hao Dali (played by Lü Liping) tells the wrenching story of losing her 3-year old son during the journey to Chengdu. Her loss echoes that suffered by millions of mothers who sacrifice their children to the nation during times of war. Now she wanders as if in a daze, an IV bottle tethered to her arm, dripping relief for an unnamed illness. She seems unable to comprehend the factory's demise, preferring to watch old war movies on television in her dormitory room.

If the workers appear to be lost souls, their children display the self-centered individualism and entrepreneurial spirit of a new generation. Zhao Gang has no interest in factory work; he apprenticed in a factory in the Northeast and was dismayed by the mind-numbing drudgery of the work. Su Na (played by Zhao Tao) is a professional shopper – a heroine of consumption – who makes monthly trips to Hong Kong to buy luxury goods for wealthy Chengdu women. We are introduced to her as she is applying lip gloss in a scene that immediately follows a long shot of the final demolition of Factory 420. "There are many rich people in Chengdu, so I can make money easily!" she tells us. She drives a sporty new Volkswagen Beetle and wears Pierre Cardin sweaters. She confesses that she does not visit her aging parents much; their house has an atmosphere of failure and depression. Her father, since retiring from the factory, is permanently grumpy. Recounting a time when she spied her mother at work on the shop floor, Su Na tells a story similar to Zhao Gang's: she is stunned by the soul-crushing toil of factory labor. "I had never felt that sad in my whole life," she exclaims tearfully as she recalls seeing her mother in her shapeless blue factory uniform. "I didn't even know if it was a man or a woman as I approached!" Her mother's sacrifice of her femininity is the ultimate indignity, and Su Na vows to set things right. "The thing I want most now is to make a lot of money. Lots and lots of money! I want to buy an apartment in 24 City for my parents. I know it will cost a lot. But I can do it. I'm the daughter of a worker!" In the film's final scene, we see Su Na atop Chengdu's Panda Tower, gazing down at the vast, endless city sprawling beneath her.

While seeming to celebrate the new middle class world of urban China, *24 City* offers a slightly off-beat, surreal, and ultimately ambivalent take on both China's urban future *and* its industrial past. Writing in *Cineaste*, Kevin Lee (2009) called it "an oral history project transformed into performance art." Jia's film, it turns out, is full of subtle reminders that China's urban common sense is in fact fragmented and incoherent, and that the memories embedded within the built environment remain up for grabs, even as that environment faces the onslaught of the wrecking ball. The documentary-style truth-telling nature of the film is unsettled by Jia's inclusion of well-known actors playing the roles of some of the workers and their children. The most jarring of these is Joan Chen's appearance as a worker knick-named "Little Flower" because she looks like the heroine in the film *Little Flower*, starring Joan Chen! "After a while my real name was known to very few people," she sighs, without a hint of irony. The audience is meant to get the joke. Jia follows this with scenes from the actual film *Little Flower*, and a much younger Joan Chen. "Sadness grips her heart" reads the caption in the older film, followed by an inter-title screen of text from the *Dream of the Red Chamber* poem, "Burial of Flowers." But this "sadness" can only be viewed as something of a tongue-in-cheek gesture; Joan Chen is giving us a wink, as Clifford Geertz might say.

Jia's sentiment, noted at the beginning of this paper, about the "chaos" of reality, seems to disrupt any possibility in *24 City* of a smooth and ordered narrative of transition and transformation in which post-industrial consumerism replaces an earlier, more meaningful time of belief in larger ideals and selfless sacrifice to the nation. Jia seems unsure if either the industrial or the urban revolutions are what they claim to be. The film opens, for example, with scenes of the heroes of China's Mao-era ideology – the proletarianized industrial workers – performing what can only be viewed as menial toil; these are not so much socialist titans of humanity as near-lifeless extensions of the hulking machines they operate. Factory labor has seemingly done little but produce damaged, worn-out machine-parts that were once humans. Indeed, in interview after interview, Jia has expressed his belief in individual expression and creativity, and how these were crushed by an ideological commitment to industrial labor under Mao. And yet, *24 City* is nevertheless framed as a paean to the glorious worker who is being displaced by a new generation of selfish brats.

Nor is the new urban future all that clear. Throughout the film, the 24 City development itself remains obscure. It is revealed to us, briefly, as one of the ubiquitous display models one sees in real estate offices throughout urban China today. Yomi Braester (2014) reads this ephemeral appearance of a maquette-version of China's urban future as a critique of the utopianism that has been invested in shiny new cities built from scratch. That it surely is, but Jia's film offers not so much a sustained critique of ideological urbanism as a wary (and weary) snort at its pretensions.

The city as creation

If we can view the 24 City development as but one version of the new town fever that swept across China since the late 1990s, then Jia's film offers a broader critique not of urbanism *per se*, but of its utopian claims. Jia's city is, like Gramsci's common sense, unreadable as anything but a text of fragments, allusions, and incoherencies.

The new town's claims of completeness are, ultimately, ideological manifestations of a utopian myth that now prevails in China in much the same way as the myth of an industry-fueled socialist modernization colonized the ideals of a previous generation.

For Lefebvre, there was a similar sense of the "unreal" to the new town. In his essay, "Notes on the New Town" (1995), he refers to the "historical city" as an organic entity. He compares it to a seashell, a living creature that has slowly secreted its structure over the years. Post-industrial capitalist society is typically built upon the ruins of this city, but those ruins give planners fits. Built next to the decaying corpse of the historical city, then, the new town solves this problem and offers an alternative space for rationally planned, technocratic, and scientific urbanism. The new town is, as Nielson and Simone (2017, 129) argue, a space of urban redemption, reviving urbanism's promise of starting anew: "This redemption cannot be envisioned through the rehabilitation of the existing city and so must be realized in the construction of a parallel city, where a future may be offered unhindered by the past."

The new town, Lefebvre argued, was an ideological edifice. Its primary function lay in colonizing everyday life for capitalism and the modern state. And its effect on everyday life was anesthetizing; the new town was "a world of unredeemable boredom" (1995, 119). The spatial legibility of the new town, its lack of ambiguity, produced a uniquely modern form of malaise. Lefebvre's new town was a contemporary, capitalist manifestation of what Bruno Fortier called the "creation city." As discussed by Doevendans and Schram (2005), the creation city was one of Fortier's three urban archetypes, the other two being the "accumulation city" and the classical Greek, or "mixed," city. The accumulation city – a city that grew and evolved over time without a conscious plan (that is, Lefebvre's seashell-like historical city) – served as a "negative model" for which modern urban planning should provide better solutions. Those solutions were epitomized in the creation city, a utopian city built tabula rasa, like ceremonial cities of ancient Egypt or contemporary political capitals like Brasilia or Canberra. Three principles inform the creation city: that the city is built exactly as it is designed and conceived (its built environment is purely ideological); that the city is purely geometrical (it displays a spatial purity); and that the city is "de-territorial" with no relationship to its location (that is, built tabula rasa). While the creation city expresses long-held utopian longings, and Judeo-Christian aspirations of creating heaven on earth (Hayden 1979), it comes to be imbued with modernity's need for order and its preference for separation and discrete spatial form.

Doevendans and Schram focus their discussion of the creation city less on a postmodern rejection of its utopian and rational pretensions than on the tension that continues to exist between the creation city and the accumulation city as competing urban paradigms. If such a tension is recognized by Lefebvre, it remains peripheral to his conviction that capitalism has colonized the creation city and turned it into a wretched machine. But for Doevendans and Schram (2005, 41), the "discord of modernity" should be understood as a *productive* tension: "The contemporary city could then be seen as a city that oscillates dialectically between these two paradigms, very much in accordance with Fortier's archetype of the Greek city, a city planned and unplanned." We might restate this tension as one between the urban ideology, on the one hand, and the process of urbanization, on the other.

China's new towns thus suggest a related set of assumptions to Fortier's creation city. They also share the machinic, abstract, even "boring" qualities of Lefebvre's new town (Shepard 2015; Sorace and Hurst 2015). Fixated on large-scale abstract geometric patterns (particularly in the layout of government buildings, CBD's and central plazas), these new zones and districts are typically promoted as eco-city and smart-city models of sustainable urbanism (Chien 2013; Williams 2017; Roggeveen and Arnold 2017; Bonino et al. 2019). Built tabula rasa next to the chaotic and polluted urban legacies of China's "producer cities" of socialist industrialism, they present themselves as an urbanism of redemption. They also present themselves as a tabula rasa for the Ozymandian arrogance of global architecture and design firms such as Kohn Pederson Fox (Changsha's Meixi Lake), Adrian Smith + Gordon Gill (Chengdu's Tianfu New District), Arup (Hangzhou's Cloud Valley), and MLA+ (Guiyang's Gui'an New Area). It is ironic that such firms are now finding eager clients in China, after the new town idea was largely discredited in the West (Provoost 2014). While schools of architecture and planning teach students that comprehensive top-down planning doesn't work (emphasizing instead acupuncture projects and bottom-up politics), municipal governments in China are proceeding on exactly this principle. The promotion of these new towns is typically couched in the language of redemption and starting anew: the creation city. Here's Nanhui New City architect Fanny Hoffman-Loss, for example: "With a completely new development you can try to form a new ideal city. Meaning that you also form a kind of ideal society without any history behind it" (cited in Shepard 2015, 116).

Of all of China's new urban zones, Shenzhen has embraced the creation city paradigm most explicitly. It presents itself (in museum narratives, promotional videos, and planning preambles) as a "city without history," an instant city (the birthplace of "Shenzhen speed") built tabula rasa, a future-leaning city, and a testament to the inevitability of modernity. Shenzhen's vision of its future self, as Mary Ann O'Donnell (2018) has pointed out, is decidedly professional, gentrified, and "middle class." Yet while Shenzhen likes to think of itself as a creation city, it is in fact a manifestation of the *tension* between the creation city and the accumulation city, between ideology and urbanization. Shenzhen, in fact, displays not only the futuristic and utopian pretensions of the new town eco-city, but also the chaotic – yet somehow effective – combination of the planned and unplanned city. Shenzhen, in other words, narrates the tension underlying *24 City*: a struggle over memory in an ever-changing landscape overseen by a state insistent on one particular story of that landscape's past (or lack thereof).

O'Donnell, Wong, and Bach (2017, 3) argue that Shenzhen's status as a new town model did not emerge as the logical result of the coherent policies and plans one would expect in a creation city (which is built as conceived). "In fact, the model emerged out of a period of illicit (and often outright politically unapproved) experimentation, which itself allowed for a series of exceptions to the model's own policies. These results were later cannily reinscribed ex post facto into officially sanctioned (and newly invented) narratives, and eagerly extended into other cities, industries, and continents." That illicit experimentation was in part due to the agency of villagers on whose land the city was built and who played a fundamental role in shaping urbanization in the region. It was also due to the government's favored mode of sketching out broad contours of objectives and targets while leaving the details for locals to sort out. The result is a city that matches Jia Zhangke's vision of an urbanization both virtual and real, where ideology and chaos reign in equal portions, deriving energy from each other.

Shenzhen's instant city narrative marks an ideological ordering imposed upon the chaotic and vernacular pattern of town and village-based urbanization in the Pearl River Delta, a region where Ren's (2013) "urbanization from below" has proceeded apace despite the state's shift toward "city-centered urbanization." What has been observed as the "generic" quality of urbanization here expresses the Western architect and planner's fascination with the region's decidedly practical, infrastructural, and seemingly unplanned urbanization – what Koolhaas (1995) referred to as urbanization without urbanity. Nielson and Simone (2017) also discuss generic cities, but more explicitly in reference to the new town phenomenon. For them, new towns built as "twins" of the historic cities they seek to redeem are "generic" in the sense of being formulaic in their engineering of urban life; they are modular forms of a "limited universalism", substitutable as ideal types, rather than as clones of their original twin. Shenzhen might thus be viewed as the "generic" twin of Hong Kong. Through their absolute detachment from existing cities, such generic cities promise the liberation of democratic livability (just as O'Donnell 2018 argues Shenzhen's planners envision a future city where all citizens can afford expensive coffee and can have direct access to their government). But, for Nielson and Simone, these spatial forms might more provocatively be read as a "diagram of a mechanism of power reduced to its ideal form" (Foucault 1975, 205). In other words, the generic city is a highly ideological city.

For Koolhaas, the generic city "is all that remains of what used to be the city" (1995, 1252). The instant quality of the Shenzhen narrative – the city that emerges from nothing – is a quality of the generic city. "All Generic Cities issue from the tabula rasa; if there was nothing, now they are there; if there was something, they have replaced it. They must, otherwise they would be historic" (p. 1253). The generic city, then, might be seen as part of the ideological narrative that seeks to obscure the actual tension between creation & accumulation city, between the planned & unplanned, the formal & informal, official & vernacular. The generic city is a harmonious city. It thrives in cozy relationship with authoritarian states and private corporate capital undisturbed by the chaos and potential threat of the public sphere.

> In the Generic City, people are not only more beautiful than their peers, they are also reputed to be more even-tempered, less anxious about work, less hostile, more pleasant – proof, in other words, that there *is* a connection between architecture and behavior, that the city can make better people through as yet unidentified methods (pp. 1262-63).

This is typical tongue-in-cheek hyperbole for Koolhaas, for whom the generic city is perhaps little more than a clever idea conjured to poke fun at the utopian pretensions of formal urbanism. If so, then Koolhaas is simply treating China as both a canvas for his own professional ambitions as a design brand, and as a foil for idle theorizing. While others have found in the generic city a nod toward the in-your-face anarchism of China's *shanzhai* knock-off culture (Kloet and Scheen 2013), Koolhaas's idea in fact bears only vague resemblance to the on-the-ground realities of urban development in China. But Koolhaas is right to suggest that the dominant form of urbanism in China is a generic, ad-hoc city that is in fact anti-design. Notwithstanding the arrogance of firms like KPF and MLA+, who promise an instant identity for instant cities based on an iconic skyline, a technological innovation and a recognizable brand, the "character" of the instant city is overpowered by urbanization itself.

What the generic city idea as wielded by Koolhaas profoundly misunderstands, though, is that the "archipelago urbanism" (Roggeveen and Arnold 2017) of the Pearl River Delta is not the result of tabula rasa city building but of particular histories of territorial administration, state population and production management, and village identities that date back, in some cases, nearly a millennium. These coalesce in processes of urbanization that have produced specific lived experiences and memories. Shenzhen, then, is hardly a city without history, despite its ideological preference to be viewed as such. And Koolhaas's generic city merely enables that ideological rendering and helps erase the memories and geographies that produced the infrastructural urbanism of the Pearl River Delta. Braester (2014, 66) agrees, arguing that Koolhaas "plays into the hands of administrators hell-bent on rapid hyper-urbanisation." It is important to recognize the memories and geographies that Koolhaas conveniently misses if we are to see the Pearl River Delta as anything other than a "model" of a new kind of urbanism. It is, rather, the extension of an infrastructural logic that has linked communities in a web of connectivities for generations.

Shenzhen emerged not simply as a single SEZ, but as a cluster of exceptional zones, built up around dispersed nodes of manufacturing (Chen 2017). These nodes were themselves village-based, the spatial outcome of a bargain between villagers and zone leaders who, in exchange for the expropriation of agricultural land, allowed villagers to retain sizable parcels of development land upon which to build export processing plants, often in partnership with Hong Kong investors, many of whom were themselves natives of these very villages. After two decades of relentless and haphazard development, Shenzhen was a city without a center, an archipelago of vernacular urbanisms. In this pattern, villages were the "starting point of urbanization, and a key element in the process of urban change that is markedly different from the dominant Western model" (Cenzatti 2014, 9). Urbanization in Shenzhen, and throughout the Pearl River Delta, "originated in the industrialization and densification of villages and towns, rather than from the expansion of the central city" (ibid.). In many respects the generic city is not a city at all, but an amalgamation of urbanized villages and towns. That much was admitted to me by a Dongguan city planner in 2017 when he said, "Actually, Dongguan doesn't exist." Meaning it doesn't exist *as a city*, despite its bulging population of nearly 10 million and its leading role among China's export manufacturing regions. While the generic city landscape of the PRD that inspired Koolhaas is certainly anti-design and ad hoc, then, it is not tabula rasa. Built without design, from the infrastructures of haphazard rural urbanization, the generic city is in many ways the *opposite* of the creation city. That fact has required that a creation city be imposed upon it *ex post facto*. The resulting blend of formal CBDs, with their geometric layouts and iconic skylines, and informal urbanized villages, constitutes the creation city-accumulation city tension that energizes urbanization in China today.

The focus on Shenzhen as a "city without history" serves the urban ideology's need to erase much of the recent urban past in China, as part of a project of reinventing the city as a space of civilization and state-led progress. Shenzhen's iconic birth – with Deng Xiaoping as midwife – matches China's ideological shift away from the coherent narrative of class struggle and toward a more diffuse embrace of the "stability" of urban consumerism. It offers a spatial materialization that replaces the industrial work unit – Factory 420 – with the commercial real estate development. (Is it any accident, then, that China's largest commercial real estate developer, Vanke, originated in Shenzhen?) Shenzhen is a built landscape that reflects the state's desire to shift the message from revolution to urbanism.

But, again, Shenzhen is a messy ideological device. Built upon a landscape of urbanized villages housing migrants from all over China in tightly packed "handshake" tenements, Shenzhen is a space in which the urban "common sense" means many different things. There is the dominant narrative of "Shenzhen speed," civilization, and modernity, a narrative that has been bent on "the elimination of the villages" (Bach 2010, 426). Urban villages were, after all, reminders of a history expunged from the record, an anachronistic drag on the city's futuristic sensibilities, a persistent legacy of what has been called "perverse" urbanization. They provide the "lump in China's urban throat" (Bach 2010, 447), always there as a testament to urbanization as a social process in tension with the city as an ideological device. And there are also competing narratives emerging from the villagers themselves. Villages like Huanggang, Shuiwei, and Xiasha have been increasingly assertive in building counter-ideologies as they reconstruct their village centers as spectacles of local tradition and Confucian patriarchy, with museums, symbolic monuments, and central plazas. These counter-narratives purposely blur the rural-urban distinction that lies at the heart of China's urban ideology. As Smith (2014) argues, in the case of Huanggang's redevelopment, "By invoking what it sees as the best of both the urban (spectacle, centrality, commerce) and the rural (filiality, consanguinity, memory), Huanggang seeks to fashion a new identity that transcends the two." As Guo and Herrmann-Pillath (2018) have pointed out in their work on Shenzhen's "ritual economy," traditional lineages were always a mix of rural and urban. While the urban ideology would expect them to wither away, the process of urbanization, at least in parts of southern China, has not worked out that way. At the same time, Shenzhen is seeing an increasingly robust heritage preservation movement, spearheaded less by villagers than by local intellectuals seeking to redefine Shenzhen identity away from the city without history and toward the mixed migrant and entrepreneurial heritage of the urbanized villages and their migrant residents (O'Donnell 2019).

The city is open

Jia Zhangke has said that "remembering history is no longer the exclusive right of the government." This sentiment is reflected in the dynamic urban village landscape of blurred distinctions that continues to trouble Shenzhen as an ideological project. This paper has suggested that we consider urban China as a tension between ideology and urbanization, and that the urban ideology offers an analytical lens that foregrounds the historical continuities of the party-state apparatus as well as the theoretical insights of Lefebvre's dialectical approach to urbanization as a social process. This does not mean, however, that such an approach offers up China as a case study in Lefebvre's urban revolution (see Tang 2014). Lefebvre's critique of capitalism qua urbanization was itself historically and geographically situated, while his apparent belief in the inevitability of urbanization vis-à-vis the countryside is based on an assumption of a fundamental urban-rural antagonism that, although reproduced by China's urban ideology, bears little resemblance to the history of social relations between city and countryside in China.

Thus, this paper has also suggested that a focus on urban ideology leaves us, counterintuitively perhaps, with an open approach to Chinese urbanization. Scholars looking at new towns in China beyond their symbolic appropriation as "ghost cities" or bubbles of real estate speculation and municipal land grabbing (all of which they indeed are), are also seeing this kind of openness in the landscape. Sampieri (2019) tells us that "the city is available", an incomplete and undecidable space awash in possibilities. Governa (2019) finds in China's new towns an image of Ananya Roy's (2009) "21st-Century metropolis" in which "margins become centers, centers become frontiers, regions become cities." She suggests that new towns might qualify as a "constitutive-outside" of the urban, an always incomplete process of becoming urban. Meanwhile, Tang (2015, 726) has documented the ways that urbanization has produced an administrative space that is "not rural but not urban … situated between the common practices of grassroots governance in the cities and the countryside."

Despite the ideological work invested in them, in other words, China's new towns and "instant cities" are in fact not all that new or instant; they are in fact outcomes of urbanization as much as messengers of the urban ideology. They are spaces where "rural and urban merge" (Governa 2019, 224), where the meaning of what is urban, suburban, rural or non-urban are constantly redefined. Finally, they are spaces where memory and history struggle to define one's reality as the city is constantly thrown into chaos, reduced to rubble, and built anew.

Notes

1. The *hukou* system has been the primary institution enforcing the rural-urban dichotomy in China (see Chan and Wei in this issue (2019)). That it no longer recognizes the categories of "urban" and "rural" (only "agricultural" and "non-agricultural") suggests a shift from one of balancing – ideologically – the urban and rural (or even favoring the latter), to one of "the urban" as the unstated social norm.
2. What I'm calling a "new city" here is meant to encompass a range of administrative and urban planning designations in Chinese. These include *xincheng* 新城, or "new town", *xinqu* 新区, or "new area", and *tequ* 特区 "special zone".

Acknowledgments

This paper has been presented, in many different iterations, to audiences at Cornell University, Simon's Rock College Great Barrington, Southern University of Science and Technology Shenzhen, Guizhou Minzu University Guiyang, Hong Kong Baptist University, and University College London. Valuable criticism and suggestions have been offered in all cases. In addition, I'd like to thank Neil Brenner, Carolyn Cartier, Kam Wing Chan, Robin McNeal, Jennifer Robinson, Wing Shing Tang, and Tammy Wong for thoughtful feedback on earlier drafts of the paper.

Disclosure statement

No potential conflict of interest was reported by the author.

References

Bach, Jonathan. 2010. "'They Come in Peasants and Leave Citizens': Urban Villages and the Making of Shenzhen, China." *Cultural Anthropology* 25 (3): 421–458.

Bach, Jonathan. 2019. "What Kind of Model? Thinking about the Special Economic Zone and the Socialist City." *Made in China* 2: 72–77.

Bonino, Michele, Francesca, Governa, Maria Paola Repellino, and Angelo Sampieri, eds. 2019. *The City after Chinese New Towns: Spaces and Imaginaries from Contemporary Urban China.* Basel: Birkhäser.

Boyer, M. Christine. 1994. *The City of Collective Memory: Its Historical Imagery and Architectural Entertainments*. Cambridge, MA: MIT Press.

Braester, Yomi. 2014. "The Architecture of Utopia: From Rem Koolhaas' Scale Models to RMB City." In *Spectacle and the City: Chinese Urbanites in Art and Popular Culture*, edited by J. de Kloet and L. Scheen, 61–75. Amsterdam: University of Amsterdam Press.

Bray, David. 2005. *Social Space and Governance in Urban China: The Danwei System from Origins to Reform*. Stanford: Stanford University Press.

Brenner, Neil, and Christian Schmid. 2011. "Planetary Urbanization." In *Urban Constellations*, edited by M. Gandy, 10–13. Berlin: Jovis.

Brenner, Neil, ed. 2014. *Implosions/Explosions: Towards a Study of Planetary Urbanization*. Berlin: Jovis.

Buckingham, Will. 2017. "Uncorking the Neoliberal Bottle: Neoliberal Critique and Urban Change in China." *Eurasian Geography and Economics* 58 (3): 297–315. doi:10.1080/15387216.2017.1350196.

Cartier, Carolyn. 2009. "Production/consumption and the Chinese City/region: Cultural Political Economy and the Feminist Diamond Ring." *Urban Geography* 30 (4): 368–390. doi:10.2747/0272-3638.30.4.368.

Cartier, Carolyn. 2015. "Territorial Urbanization and the Party-state in China." *Territory, Politics, Governance* 3 (3): 294–320. doi:10.1080/21622671.2015.1005125.

Cartier, Carolyn. 2019. "Paradigm, Analog, and Exemplary Cities in China: The Capitalist Aesthetic and the Loss of Space." *Eurasian Geography and Economics* 60 (4).

Cartier, Carolyn, and Luigi Tomba. 2014. "Urban/ideology and Contemporary China: Toward an Agenda for Studying China's Urban Condition." Australian Centre on China in the World, Australian National University (16 January).

Cenzatti, Marco. 2014. "The City between the Villages." In *Villages in the City: A Guide to South China's Informal Settlements*, edited by A. Stefan, 9–16. Hong Kong: University of Hong Kong Press.

Chan, Kam Wing. 1994. *Cities with Invisible Walls: Reinterpreting Urbanization in Post-1949 China*. Hong Kong: Oxford University Press.

Chan, Kam Wing. 2007. "Misconceptions and Complexities in the Study of China's Cities: Definitions, Statistics, and Implications." *Eurasian Geography and Economics* 48 (4): 383–412. doi:10.2747/1538-7216.48.4.383.

Chan, Kam Wing, and Yanning Wei. 2019. "Two Systems in One Country: The Origin, Functions, and Mechanisms of the Rural-Urban Dual System in China." *Eurasian Geography and Economics* 60 (4).

Chen, Ting. 2017. *A State beyond the State: Shenzhen and the Transformation of Urban China*. Rotterdam: nai010.

Chien, Shiuh-Shen. 2013. "Chinese Eco-cities: A Perspective of Land-speculation-oriented Local Entrepreneurialism." *China Information* 27 (2): 173–196. doi:10.1177/0920203X13485702.

Doevendans, Kees, and Anne Schram. 2005. "Creation/accumulation City." *Theory, Culture & Society* 22 (2): 29–43. doi:10.1177/0263276405051664.

FloraCruz, Jaime. 1994. "The Crisis of Ideology in China." *The Brown Journal of World Affairs* 2 (1): 169–174.

Foucault, Michel. 1975. *Discipline & Punish: The Birth of the Prison*. New York: Vintage.

Gaubatz, Piper. 1996. *Beyond the Great Wall: Urban Form and Transformations on Chinese Frontiers*. Stanford, CA: Stanford University Press.

Governa, Francesca. 2019. "Scaling up and Scaling Out. New Towns and 'the Standpoint of an Absence." In *The City after Chinese New Towns: Spaces and Imaginaries from*

Contemporary Urban China, edited by M. Bonino, F. Governa, M. P. Repellino, and A. Sampieri, 215–226. Basel: Birkhäser.

Gramsci, Antonio. 1971. *Selections from the Prison Notebooks*. New York: International.

Guo, Man, and Carsten Herrmann-Pillath. 2018. "'iClan': The Chinese Ritual Economy in the 21st Century." Accessed 22 April. https://ssrn.com/abstract=3166765

Harvey, David. 2005. *A Brief History of Neoliberalism*. Oxford: Oxford University Press.

Harvey, David. 2016. "Senior Loeb Scholar Lecture at Harvard Graduate School of Design." 28 March. Accessed 20 February 2017. https://www.youtube.com/watch?v=pm_UgX–ef8

Hayden, Dolores. 1979. *Seven American Utopias: The Architecture of Communitarian Socialism, 1790–1975*. Cambridge: MIT Press.

Howard, Pat. 1988. *Breaking the Iron Rice Bowl; Prospects for Socialism in China's Countryside*. Armonk, NY: M.E. Sharpe.

Hsing, You-tien. 2010. *The Great Urban Transformation: Politics of Land and Property in China*. Oxford: Oxford University Press.

Johnson, Ian. 2013. "China's Great Uprooting: Moving 250 Million into Cities." *The New York Times* 15 June. http://www.nytimes.com/2013/06/16/world/asia/chinas-great-uprooting-moving-250-million-into-cities.html

Kloet, Jeroen de, and Lena Scheen. 2013. "Pudong: The *Shanzhai* Global City." *European Journal of Cultural Studies* 16 (6): 692–709. doi:10.1177/1367549413497697.

Koolhaas, Rem. 1995. "The Generic City." *S,M,X,XL*, 1239–1264. New York: Monacelli Press.

Le Monde. 2011. "La Révolution Chinoise, Selon Jia Zhang-ke." 18 January. Accessed 1 August 2019. https://www.lemonde.fr/cinema/article/2011/01/18/la-revolution-chinoise-selon-jia-zhang-ke_1467186_3476.html

Lee, Kevin. 2009. "Review of *24 City*." *Cineaste* 34 (4): 44–46. Fall.

Lefebvre, Henri. 1995. "Notes on the New Town." *Introduction to Modernity: Twelve Preludes, September 1959-May 1961*, J. Moore, trans, 116–126. London & New York: Verso.

Lefebvre, Henri. 2003. *The Urban Revolution*. R. Bononno, trans. Minneapolis: University of Minnesota Press.

Lim, Louisa. 2014. *The People's Republic of Amnesia: Tian'anmen Revisited*. Oxford: Oxford University Press.

Lin, George. 2011. "Territorialization of State Power through Land Development in Southern China." *Cross-Currents: East Asian History and Culture Review* 1 (1). http://cross-currents.berkeley.edu/sites/default/files/e-journal/articles/g._s._lin_0.pdf

Liu, Geng. 2014. "Chengdu Xin Nongcun Shijian: Meiyou Chanye Zhicheng Jiu Yici Xing Penjing" [chengdu's New Countryside Experience: Without the Support of Property, It's Just a Potted Plant]." *Liaowang Dongfang [Oriental Outlook]* 33, August 25.

Looney, Kristen. 2012. "The Rural Developmental State: Modernization Campaigns and Peasant Politics in China, Taiwan, and South Korea." PhD Dissertation, Harvard University, Cambridge, MA. doi:10.1094/PDIS-11-11-0999-PDN.

Ma, Lawrence, and Fulong Wu, eds. 2005. *Restructuring the Chinese City: Changing Society, Economy, and Space*. London and New York: Routledge.

Nielson, Morten, and AbdouMaliq Simone. 2017. "The Generic City: Examples from Jakarta, Indonesia, and Maputo, Mozambique." In *Infrastructures and Social Complexity: A Companion*, edited by P. Harvey, C. B. Jensen, and A. Morita, 128–140. London & New York: Routledge.

Nora, Pierre, and Lawrence Kritzman. 1998. *Realms of Memory: Rethinking the French past (vol. 1)*. A. Goldhammer trans. New York: Columbia University Press.

O'Donnell, Mary Ann. 2018. "Excavating the Future in Shenzhen." In *Urban Asias: Essays on Futurity past and Present*, edited by T. Bunnell and D. Goh, 247–262. Berlin: Jovis.

O'Donnell, Mary Ann. 2019. "The New Companion to Urban Design." In *Heart of Shenzhen: The Movement to Preserve 'ancient' Hubei Village*, edited by T. Bannerjee and A. Loukaitou-Sideris, 480–493. London & New York: Routledge.

O'Donnell, Mary Ann, Winnie Wong, and Jonathan Bach, eds. 2017. *Learning from Shenzhen: China's Post-Mao Experiment from Special Zone to Model City*. Chicago: University of Chicago Press.

Provoost, Michelle. 2014. "Why Build a New Town?" In *Volume 34: City in a Box*, edited by A. Oosterman, 5–7. Amsterdam: ARCHIS.

Ren, Xuefei. 2013. *Urban China*. Cambridge: Polity.

Roggeveen, Daan, and Frances Arnold, eds. 2017. *Progress and Prosperity: The Chinese City as Global Urban Model*. Rotterdam: nai010.

Roy, Ananya. 2009. "The 21st-century Metropolis: New Geographies of Theory." *Regional Studies* 43 (6): 819–830. doi:10.1080/00343400701809665.

Sampieri, Angelo. 2019. "The City Is Available: Chinese New Towns as a Backup Space." In *The City after Chinese New Towns: Spaces and Imaginaries from Contemporary Urban China*, edited by M. Bonino, F. Governa, M. P. Repellino, and A. Sampieri, 204–214. Basel: Birkhäser.

Schmid, Christian, Lukasz. Stanek and Akos Moravanszky. 2014. "Theory, Not Method - Thinking with Lefebvre." In *Urban Revolution Now: Henri Lefebvre in Social Research and Architecture*, edited by L. Stanek, C. Schmid, and A. Moravanszky, 1–24. Farnham: Ashgate.

Schurmann, Franz. 1968. *Ideology and Organization in Communist China*. 2nd ed. Berkeley: University of California Press.

Shepard, Wade. 2015. *Ghost Cities of China*. London: Zed Books.

Smil, Vaclav. 2014. *Making the Modern World: Materials and Dematerialization*. Chichester: Wiley.

Smith, Nick. 2014. "City-in-the-village: Huanggang and China's Urban Renewal." In *Villages in the City: A Guide to South China's Informal Settlements*, edited by A. Stefan, 29–40. Hong Kong: University of Hong Kong Press.

Sorace, Christian, and William Hurst. 2015. "China's Phantom Urbanisation and the Pathology of Ghost Cities." *Journal of Contemporary Asia* 46 (2): 304–322. doi:10.1080/00472336.2015.1115532.

Swanson, Ana. 2015. "How China Used More Cement in 3 Years than the US Did in the Entire 20th Century." *The Washington Post*, 24 March. Accessed 20 February 2019. https://www.washingtonpost.com/news/wonk/wp/2015/03/24/how-china-used-more-cement-in-3-years-than-the-u-s-did-in-the-entire-20th-century/

Tang, Beibei. 2015. "'Not Rural but Not Urban': Community Governance in China's Urban Villages." *The China Quarterly* 223: 724–744. doi:10.1017/S0305741015000843.

Tang, Wing Shing. 2014. "Where Lefebvre Meets the East: Urbanization in Hong Kong." In *Urban Revolution Now: Henri Lefebvre in Social Research and Architecture*, edited by L. Stanek, C. Schmid, and A. Moravanszky, 71–91. Farnham: Ashgate.

Wachsmuth, David. 2014. "City as Ideology." In *Implosions/Explosions: Towards a Study of Planetary Urbanization*, edited by N. Brenner, 353–371. Berlin: Jovis.

Wilczak, Jessica. 2017. "Making the Countryside More like the Countryside? Rural Planning and Metropolitan Visions in Post-quake Chengdu." *Geoforum* 78: 110–118. doi:10.1016/j.geoforum.2016.10.007.

Williams, Austin. 2017. *China's Urban Revolution: Understanding Chinese Eco-Cities*. London: Bloomsbury.

Wu, Fulong, ed. 2009. *China's Emerging Cities: The Making of New Urbanism*. London and New York: Routledge.

Wu, Shu-chin. 2011. "Time, History, and Memory in Jia Zhangke's *24 City*." *Film Criticism* 36 (1): 3–23.

Wu, Weiping, and Piper Gaubats. 2013. *The Chinese City*. London & New York: Routledge.

Zhang, Yingjin. 2010. *Cinema, Space, and Polylocality in a Globalizing China*. Honolulu: University of Hawai'i Press.

Two systems in one country: the origin, functions, and mechanisms of the rural-urban dual system in China

Kam Wing Chan and Yanning Wei

ABSTRACT

This paper takes an interdisciplinary perspective to focus on the role played by the rural-urban dual system in China's urbanization and migration. We explain its establishment in China in the 1950s and its genealogy in the industrialization strategy adopted in the USSR in the late 1920s and early 1930s. In the reform era, the system has been refashioned to suit a new set of demands of the state industrialization strategy to cater for the global economy. We analyze the mechanisms of the dual system, and how it has helped generate low-cost migrant labor for urban employers and siphon off revenues for urban governments by expropriating rural land. We also argue that there are fundamental differences between the dual system and the common "dualism," based on the Lewis dual-sector model, as observed in many developing countries. The plight of children of migrants is used to illustrate the main points made in the paper. Our research challenges the prevailing perspectives in Chinese geography and urban studies, many of which are based on market-driven neoclassical economics or global neoliberalism, often ignoring China's historical and institutional contexts.

Introduction

Closely associated with urbanization and rural-urban migration, rural-urban relations are a critical aspect in the process of development, especially in the modernization theories of industrialization and urbanization (Moore 1984). Seminal works on development, such as Lewis (1954), Huntington (1968), Todaro (1969), and Lipton (1977) have underlined this point. Indeed, the "agrarian question" and the related notion of proletarianization are also central in the Marxist theory of the development of capitalism (Bideleux 1985). Of course, the coercive transfer of the rural surplus from agriculture is a hallmark of the socialist[1] industrialization model first practiced under Stalin in the USSR. China under Mao also followed this approach, relying heavily on exploiting agriculture in its industrialization.[2] With a large rural population to begin with,

This article has been republished with minor changes. These changes do not impact the academic content of the article.

rural-urban relations in China have become even more central in that process (Lardy 1985; Selden 1993). As a result, the "*sannong*" ("three rural"[3]) issues were major perennial problems in recent decades. Several times in the post-1949 era, agriculture was squeezed to the point of total collapse, as in the famines in the early 1960s, or on the brink, as in the mid- and late 1970s. Since the rural sector and its population have been under recurrent stress, serious problems in the rural sector have surfaced from time to time in the reform era, even though China's recent economic growth is considered by some a "miracle" (Lin, Cai, and Li 2003). Notwithstanding China's fast industrialization of the economy in the last four decades witnessing the share of agriculture in the GDP dropping to below 10 percent, the population legally bound to the rural sector (*nongye hukou*) based on the household registration remains very large (at a whopping 60 percent of the nation's population in 2016!) – and continues to subsist at a much lower economic level than its urban counterpart (Chan 2018a). This phenomenon is rather ironic and unusual, also requiring more investigation.

Indeed it is widely noted that the social and economic cleavages between the rural and urban populations are huge in China. This paper examines China's special rural-urban dual system (*chengxiang eryuanzhi*), a system which legally, economically, and socially accords different treatments to the rural and urban populations. Understanding its historical roots and operations of this system is central to an accurate comprehension of the nature of China's post-1949 development. This specificity is also well reflected in China's rather unusual "incomplete" urbanization pattern and "semi-proletarianization" of rural-urban migrants, both of which have been fairly well studied (Chan 1994, 2018b; Zhang, LeGates, and Zhao 2016; Selden and Wu 2011; Pun and Chan 2012). As will be demonstrated, the dual system is foundational to the Stalinist/Maoist state socialist development strategy and the political-cum-economic system adopted by the Chinese Communist Party (CCP). The "two systems" has undergirded not only industrialization under Mao, but also urbanization and rural-urban dynamics in the reform era.

This omnipotent, mega dual system and its related problems, apparently based on a certain socialist political economic logic alluded to above, received a fair amount of attention in China, especially in the 1980s and 1990s, such as in the pioneering work of Guo and Liu (1990),[4] and even at recent Party Congresses and Plenums.[5] However, a systematic treatment of the mega dual system, particularly in the reform era, is still wanting. Indeed, this institution and its underlying development logic remain largely in the peripheral field of attention, if not the blind spot, of most mainstream social science literature in the West interpreting China's development and rural-urban relation and associated issues and processes, such as urbanization and migration. Perhaps this is not totally surprising given the problematic dominant paradigms that much of the Western geography and urban studies scholarship on China has relied on. As Tang (2014), Buckingham (2017), and Zhou, Lin and Zhang. (2019) have pointed out, the "master narrative of neoliberalism" is especially problematic, and this paradigm is more systematically critiqued by Cartier (2019), Oakes (2019) and Tang (2019) in this issue.[6] A large body of English literature on China's development and rural/urban issues (such as urban development, and the plights of migrant workers) in the reform era is often based on the fundamentals of economic liberalism, to the extent that the continuity of Mao's socialist institutions in the current era and the power relations of the party-state are sidelined if not ignored.[7] Many of those works are ahistorical, lacking an understanding of the continuity of the Maoist socialist practices and the Soviet roots of the Chinese institutions. Those studies assume the post-Mao reforms "as a sort-of 'ground zero' for the emergence of a new, globalized and quasi-capitalist China" (Oakes 2019). Hence, it is not too shocking to see the conflation of seemingly similar but actually quite different concepts and phenomena. For example, the Anglo-American urban design concept "urban village" for middle-class housing has been, to use Tang's (2014) words, "randomly appropriated" without taking into account the historical and institutional contexts to frame the narrative of a totally different entity: China's slum-like "*chengzhongcun*" (villages-in-the-city) – indeed, they are truly urban slums for the bottom rung of the city (Huang and Yi 2014; Yao 2014). Equally troubling, to cite one more example, there has been almost perennial and very serious problems in deciphering China's city population statistics largely because of the mix-up of different statistical concepts based on a command economy with Western ones of a market economy.[8] In this issue, Cartier (2019) has pointedly queried, "why do paradigms adopted for research on China derive more often from the experience of the urban process under capitalism rather than its legacies and transformations from the history of the planned economy under socialism?"

Building on the previous work by Chinese scholars cited above, and research in the West, notably Chan (1994), Chan (2009), Chan and Zhang (1999), Solinger (1999), Whyte (2010b), Wu (2017), and Buckingham (2017), this paper takes one more step forward to systematically study the roots and nature of the Chinese rural-urban dual system so that the Chinese development experience is accurately portrayed and understood. As will be argued below, the Chinese dual system is not just simply a rural-urban divide in the economy, as in the well-known, market-driven dualistic-sector, rural-to-urban labor migration model (Lewis 1954). Instead, the Chinese dual system is part and parcel of a far more potent and coercive force in action – a socialist state strategy to industrialize the country on the backs on the peasantry.

This paper takes an interdisciplinary China studies perspective, laying out the historical context to draw important connections in industrialization strategy between the USSR and China, and presents a set of new analyses focusing on the rural-urban divide and relations, as well as migration. Our paper is not to espouse China "exceptionalism" but challenge the prevailing market-based, especially, neoliberalism paradigms in Chinese geography, urban studies and related fields, and to put China back in the proper contexts. The next section examines the roots of the dual system, as a strategy adopted in the USSR in the late 1920s and early 1930s. Section 3 explains the establishment of the system in China in the 1950s and the major components of the state development strategy. Section 4 takes the analysis to the reform era and looks at how the system is refashioned to suit a new set of demands of the state industrialization strategy and markets. Section 5 provides a synthesis of the mechanisms of the dual system and its features, including its differences with the Lewis model in some fundamental ways. The example of China's children of migrants is used to illustrate the main argument. The conclusion summarizes our analysis.

The Soviet roots of the rural-urban dual system

The paramount goal of the Chinese state in the last several decades has been to achieve "national development" through industrialization. After the founding of the People's Republic in 1949, faced with a country that had only a small base of industry after decades-long wars, the CCP leadership opted to industrialize quickly and the Soviet Union was the ready model.[9] For Mao, the USSR was "the best teacher" in economics despite its attempt to indigenize the CCP in the 1940s (Meisner 1999; Bernstein 2014; Gao 2018). The Soviet industrialization model refers to the strategy implemented in the USSR under Stalin. The model is characterized by sustained high growth rates of industry based on massive predatory extraction of agriculture to finance industrial growth. The economic and administrative means used were collectivization and, especially, the manipulation of farm commodity prices.

In Russia, shortly after the October Revolution in 1917, the biggest challenge faced by the Bolsheviks was to deal with the free petty landed peasants that made up the great majority of the population. After the failed experience of War Communism in 1919–21, during which private trade and the use of money were eliminated, Lenin experimented with the New Economic Policy (NEP) in 1921–28, which allowed the private farm sector to grow under the "the dictatorship of the proletariat" (Soviet Communist Party) as a means to reviving the economy (Gregory and Stuart 1986).

The great industrialization debate ensued in the mid-1920s between the Party's major economists and political theorists, mainly Bukharin and Preobrazhensky. Advocating a dynamic economic equilibrium, Bukharin argued that only a "balanced, steady relationship between the interests of industry and agriculture would secure healthy economic development" (Service 2006, 260). In his view, the growth of industry could be done only through the development and support of agriculture (Boettke 1990). In *The*

Politics and Economics of The Transitional Period, he contended that "a new society cannot suddenly appear like a *deus ex machina*. Its component elements develop within the old society, ... the elements of the new society must be sought in the relations of the production of the old." Then he asked: "which aspects of the relations of production in a capitalist society can form the basis of the new production structure?" For him, it was the "maturity" of capitalist society that would serve as the stepping stone in the transitional era between a socialist society and a communist society (Bukharin 2007, 94).

Preobrazhensky, on the other hand, was more alarmed about the growth of the private economy under the NEP. He contended that a growing capitalist economy in the countryside would not help industrialize the Soviet state that was surrounded by hostile capitalist countries. He considered that since agriculture in Russia had been backward and low-efficient in comparison to that in advanced capitalist countries, it would take a long time before it could catch up and be able to support industry (Boettke 1990). He predicted that the internal and external situation that the Soviet Union faced in the 1920s would not allow any delay in rapid industrialization. Therefore, he argued that fixed capital had to be increased before the output of manufactures could be expanded and therefore it was necessary to make a commitment to heavy industry (Burdekin 1989). Specifically, in *The Law of Primitive Socialist Accumulation*, Preobrazhensky (1980) put forward that to achieve rapid industrialization, the Soviet Union needed to directly and forcibly get surplus product from the private economy, i.e., the peasantry, through a distorted pricing mechanism called "nonequivalent exchange" between industrial and agricultural commodities rather than through market. That had to be accomplished at the expense of the peasantry. The center of this debate between Preobrazhensky and Bukharin revolved around the choice of sectoral priorities (Lipton 1977; Bideleux 1985) underlying the centrality of the industrial-agricultural relation and the derivative rural-urban relation in the Soviet industrialization model.

After consolidation of political power in 1926, Stalin finally put Preobrazhensky's theory into practice in the First Five-year Plan in 1928–32. Preobrazhensky was among the earliest theorists who realized the *law of value* could be manipulated through a planned economy and resources necessary for industrialization could be quickly transferred and accumulated through an unequal exchange between industry and agriculture. But this extraction had to be done with coercion, not through the market. His design laid the foundation for the implementation of the centrally planned economy in the USSR; the strategy also required setting up separate systems of social control in the rural and urban sectors, and this was achieved by the nationwide establishment of state-controlled collective farms in the early 1930s and an internal passport system (Fitzpatrick 1996; Allen 2009). Economics aside, the decision to subordinate the interests of the peasants to those of the industry and the proletarian minority, as Bideleux (1985, Chapter 1) argues, can be also partly attributed to Marx's and Engels' rather damning views on the role of the peasantry in the industrial-capitalist transition to socialism.

In summary, to achieve such an extraction from the predominantly landed peasantry, a trinity of tools was needed. As scholars described later in the case of China, they were the compulsory procurement and monopoly of sales of farm produce, rural collectivization, and a system to control population mobility (Cai 2018). The first tool was to generate an unequal inter-sectoral (economic) exchange through eliminating the market, and the second and third served as the administrative, policing and implementation mechanisms to ensure the success of the first through control of the society, especially population mobility. An internal passport system was established to regulate residency registration (*propiska)* and many other economic, social and political activities of individuals and families (Zaslavsky and Luryi 1979). Indeed, a "passport society" was created within the USSR (Matthews 1993). This was an important step in establishing a totalitarian regime by Stalin to control the society, beginning with preventing out-migration of the peasantry. In fact, peasants were not given internal passports and could not move beyond their farms to access other employment and opportunities and services except through recruitment initiated by the state (Matthews 1993).[10] Scholars of Russia have referred to this as "the end of the free peasantry" (Conquest 1987). Others have called this practice the "enserfment of the Russian peasantry" (Smith 1968) or "revival of serfdom" (Podrabinek 2013), in a country where serfdom was practiced for centuries until 1861 and a *proposka* was in place under the Tzar until 1917. To use a common metaphor in economics to explicate the functions of the three tools described above: the first device was to handcuff the Smithian "invisible hand", controlling the exchange prices; the last two tools were to shackle the Tieboutian "feet" to prevent exit from agriculture.[11] And as will be expounded below in the Chinese case, the effect of state control went much farther beyond the narrow sphere of eliminating the market.

Establishment of the rural-urban dual system in Mao's China

As said before, the choice of the industrial-agricultural sectoral priorities was at the core of the Stalinist development strategy. The model was premised on exploitation of agriculture – in fact, relying heavily on the "tribute" from agriculture to finance rapid industrial growth (Tang 1984; Wemheuer 2014). Although Mao was aware of the deleterious effects of the Stalinist strategy on the rural sector, he nonetheless followed the Soviet approach (Levine and Rozman 1988; Wemheuer 2014). He held a hardline on the issue and harshly criticized some Party officials who were reluctant to forcibly promote the "cooperative movement" in the countryside, a precursor to full collectivization. Some officials were demoted because they failed to demonstrate a "passion for socialism" (Bo 1991, 373). As expected, the industrialization strategy carried out under Mao included all the major components of the Soviet approach (Tang 1984; Yang 1987; Naughton 2007). In particular, collectivization was implemented in the countryside in 1953–56, through which peasants lost much of their *de facto* control of their produce, labor and land.

An important institutional foundation that has made massive and predatory extraction from agriculture possible was the division of the society and economy into two rather disparate systems, commonly referred to in China as the "urban-rural dual system" (*chengxiang eryuanzhi*) (Wang 2013; Wemheuer 2014). In short, this separation enabled the different treatments of industrial and agricultural sectors, and the associated rural and urban populations. This led to a persistent dual social structure with a permanent second class of peasants. The *hukou* system and the land system were central to the enforcement of that structure.

The system of registering population and households has a long history in China. But in the pre-1949 era, the system was used primarily for collecting vital statistics tied to tax collection and conscription purposes. During times of unrest and wars, the system was employed for controlling certain groups of the population (Ge, Wu, and Cao 1997). However, the system never functioned as a comprehensive, all-embracing social and economic control mechanism in peacetime for the great majority of the peasantry. Free rural-urban mobility, though seldom used, was never forbidden in normal times (Chao 1986; Wong 2008). The gradual development of an all-encompassing version of the *hukou* system during the 1950s was new to China, and was an inevitable outcome of, or necessary part of the establishment of a command economic system in pursuit of an industrialization strategy based on the Soviet industrialization logic explained above (Wang 2013; Zhou 2017, 55–69). Accomplishing the industrialization goal also required meticulous planning and control of all macro and micro facets of society (Chan 1994, 2009; Naughton 2007).

The differential treatment of industry and agriculture (and its respective population of labor and land) lay at the core of the rural-urban dual system (Table 1). Industry, almost totally located in cities, was designated as the priority sector of the economy and was state owned. It was put under strict state management and received strong state support and protection and was the sector that generated revenues for the state. Basic state-provided social welfare and subsidies for urban workers and their families were also put in place to maintain social and political stability of the urban system, which covered about 15–16 per cent of the population in 1955. That level was maintained for almost every year until 1978 (see Chan 2009, Table 4.1).

Agriculture was the non-priority sector encompassing the rest, roughly 85 per cent of the population in the mid-1950s and had a separate, minimal welfare arrangement. As summarized in Table 1, the agricultural sector was basically excluded from state-provided welfare; agriculture's role, instead, was

Table 1. China's rural-urban dual system in Mao's era.

Economy	
Industry	Agriculture
Priority sector	Non-priority sector
State owned (including land)	Collective owned (including land)
State support and control	Self-reliance
Monopoly profits through unequal exchange with agriculture	As provider of cheap resources for the state sector
Society (based on *hukou* classification)	
Urban/"Non-agricultural" Households	Rural/"Agricultural" Households
State protection; subject to close political control	No claim to national social welfare; subject to less state control
State-provided employment, housing and welfare (such as retirement benefits)	Subsistence employment in agriculture with limited welfare provided by collectives
Restricted entry	Tied to land and agriculture; no migration allowed
High in the administrative hierarchy	Low in the administrative hierarchy

Source: adapted from Chan and Zhang (1999).

to furnish cheap raw materials (mainly food grain), labor, and capital for the state urban-industrial sector. Rural production with its labor, capital and land, was collectivized to serve the above functions, with the collectives operating basically as the state's agent. The farm population was expected to fend for itself except during time of extreme duress or emergencies. Collectivization enabled the state (through local governments) to gain control of the use of rural labor, rural land and, ultimately, grain production and distribution. With "tribute" from the rural sector, industry generated monopoly profits for the state coffers to carry out ambitious industrialization programs (Sun and Hu 1997). Most ironically, despite ongoing state propaganda touting the elimination of the rural-urban gap, investments in the rural sector and agriculture steadily declined over time in Mao's era so that the gap actually widened (Lardy 1985; Brown 2012).

The *hukou* system, which Chan (1994) calls the "invisible walls," was the necessary instrument in making this sectoral and institutional segmentation possible. Under the *hukou* system, each individual is fixated within his/her own small *hukou* administrative unit (a neighborhood in the city, or a village) at a different position in the hierarchy within the broader rural-urban division.[12] Permanent migration from one administrative unit to another was not permitted without the approval of the *hukou* authorities. Furthermore, in addition to the huge rural-urban disparities in state-provided services, the *hukou* location also determined the level of benefits that one could get. Benefits also varied from one administrative unit to another even at the same level in the hierarchy. This part of the system remains largely the same today (Cartier 2015).

As it happened in many other communist countries before, this Stalinist industrialization strategy created huge disparities between the urban-industrial and the agricultural sectors, and was bound to generate immense pressures of outflows from the countryside until mobility controls were fully in place. As China pressed forward with industrialization in the 1950s through collectivization, large numbers of peasants began to flee to the cities. Even though the freedom of migration and residence had just been freshly enshrined in the first Constitution in 1954, the state took measures to stem the flows including the designation of the population as "urban" or "rural" in 1955 and by setting up checkpoints at various major transportation nodes from 1955 through 1957 (Chan 2009). It soon became obvious that a more systematic and coercive mechanism would be needed to prevent or at least regulate these "undesirable" rural-to-urban migratory flows, called "blind flows" (*mangliu*). It was then that the *hukou* was finally codified in January 1958 and came into full play as a central component of the command system. The *hukou* system required that all internal migration be subject to approvals from the local authorities, just like the Soviet's *propiska* system. Each person's *hukou* was classified as rural or urban. It was also hereditary: newborns' classification followed that of the mother.

By immobilizing the peasantry, forcing them to tend the land at mostly subsistence levels of compensation, and excluding them from access to social welfare and many other opportunities, this approach created two very different societies – indeed, two systems in one country. Wong (2008) characterizes this change the "calcification" of the divide between the urban and rural sectors. Taking away peasants' ability to move out of the countryside meant closing the escape route from poverty, a major remedy they had in the pre-1949 times. Whyte (2010a) and Gardner (2017) call this system "socialist serfdom". This is where the Soviet/Chinese dual system differs from the typical Lewisian dualistic sector model, which is based on a free market and free migration (examined in Section 5).

The dual system in the reform era

The near collapse of the Chinese economy in the late 1970s led the post-Mao leadership to carry out a series of dramatic reforms and abandon the ideology of class struggle (Oakes 2019). Quite different from before, China began to engage in the global economy and production in 1979 and gradually adopted a series of market-oriented, "capitalist" economic measures.[13] At the same time, the economy has increasingly moved away from what Kornai (1992) called a "shortage economy" to one more responsive to the consumer. While Mao's extraction of the rural surplus focused on siphoning agricultural output at low prices when China remained largely an agricultural economy, the forms and the targets of that process used in the reform era differed significantly, as the new era of industrial production for the global economy ushered in. In the following, we analyze two major aspects of the role the dual system has played in industrialization and urbanization in the reform era.

Making cheap migrant labor for the "world's factory"

One of the most important changes was the dismantling of agricultural collectives, a major component of the Stalinist-type development model. Decollectivization unleashed a large army of hidden rural surplus labor. Despite many significant reforms in this era, the dual system, along with its major institution, the *hukou* system, was not abolished but instead gradually refashioned to serve China's new strategy of becoming the "world's factory" and more. A major change in the early 1980s was to allow peasants first to enter the cities to sell produce from their farms and, later in the mid-1980s, to seek work in the cities without getting the collectives' approval. When China's export-processing industry roared into high gear in the 1990s, the deployment of rural-*hukou* labor – in the form of "temporary migrant labor" – to the cities for the export industry became a major strategy. By the mid-1990s, rural-*hukou* labor had become the backbone of the manufacturing sector and of urban construction (Chan 2009). The voluminous migrant labor to the cities was institutionally set up as "temporary," principally via the means of the *hukou* system. These workers gradually became a prodigious army of industrial and service workers with rural *hukou* residing in cities, called "rural migrant workers" (*nongmingong*). This population grew from about 20 – 30 million in the early 1980s to about 170 million at the end of 2017 (Chan and Yang 2019). These laborers, however, were not legally considered urban workers, and were therefore not eligible for the welfare and rights that are available to urban residents. Nor were they supposed to settle in the destination. Migrant workers also face many formal and informal obstacles to access jobs other than low-skilled ones (Solinger 1999; Cai 2007). Without local *hukou* and the plentiful supply of migrant workers (until recently), they have become a huge class of super-exploitable, yet highly mobile and flexible industrial workers for China's new export economy (Lee 1998; Alexander and Chan 2004). The "temporary" nature of their urban citizenship marginalizes them, making them "in the city but not of the city." (Chan 2011).

That "not of the city" underlies the super low-cost character of Chinese migrant labor and contributes significantly to the low "China price" (Harney 2008). Not only are direct labor costs very low, but also the "indirect labor costs" – of housing the family, health care, retirement, and educating and feeding future members of the labor – are little or nil to the employer or urban government (Chan and Robert 2003). These costs of the social reproduction of labor are largely outsourced to the countryside or absorbed by the migrant family (Pun and Ren 2008; Chuang 2015). This rural "social provisioning" system for the urban (migrant) laborer and family is made possible by the dual system, which also prevents the rural-*hukou* population from exiting the rural system except to sell their labor power. In Arrighi's (2007) parlance, this type of growth is based on "accumulation without dispossession," quite different from the neoliberal approach of capitalist accumulation through dispossession, as in Harvey (2005). Selden and Wu (2011) and Pun and Ren (2008) call this process "incomplete proletarianization" or "semi-proletarianization," made possible by the *hukou* system.[14]

When the 2008 global financial crisis hit China's export sector hard, it was the migrant workers and their families in the countryside who bore the brunt: some 20 million migrant workers who had no unemployment coverage to fall back on and who had to absorb the loss by themselves and their families were laid off (Chan 2010c). The new approach of "freeing" peasant labor but under the continuing dual system through dispossession or "without dispossession," has served China's economic growth strategy of being the world's largest low-cost industrial producer very well. Effectively, the persisting dual system has enabled Chinese enterprises and the state to continue to draw labor from the countryside to work in urban areas and export-processing zones at low, rural-subsistence wage rates for a much longer time (Chan 2010b).

State expropriation of rural land

Similar to this rendition of peasant cheap urban labor on the basis of the dual system, urban governments in the reform era have figured out another important way to siphon off the rural surplus via state expropriation of rural land to generate revenues (Liu 2018). This process of "accumulation through dispossession" has gradually become a new state development strategy since the 1990s, operating largely through the system of rural collective land ownership. Collectivization in the 1950s transferred private rural land ownership into collective ownership and gave the state, if indirect, control of rural land (Lin 2009). The change also made the state the sole authority to expropriate land from rural collectives for urban use, giving it the monopoly of the windfall profits accruing from the conversion of agricultural land for nonagricultural use in urbanizing regions (Liu 2017).

During Mao's era, when industrialization and especially urbanization were at a much lower level and the demand for converting farmland to new urban-industrial land was small, the value of farmland was very low. In the absence of a land market under Mao, in fact, rural land had no real market value. However, the story became quite different as export industrialization moved apace in the reform era, which fueled rapid rise in demand for land for industrial and urban use and municipal governments turning to land as generator of revenues in the last three decades. A market for urban-industrial land also began to be gradually developed in the late 1980s when transfer of urban land use rights was legalized (Liu 2018). At the same time, the growth of an urban middle class, combined with waves of intense speculation, has driven rapid expansion of an urban real estate sector and demand for land for housing and business. Furthermore, in the wake of the global financial crisis of 2008–09, the state-directed, investment-driven development entered a new phase, creating an urban infrastructure investment boom (Tsui 2011). The new state development paradigm centers on urban construction by local governments, with financing derived from new land expropriated from the peasants being the major new source. Several campaigns have been launched to facilitate state expropriation of rural land. The most notable one, perhaps, is the "building a new socialist countryside" program from 2006, leading to mass conversion of rural land for urban uses in many provinces (Ong 2014; Chuang 2015).

While in a true market economy, the greatly increased demand for land for urban and industrial use would bestow large monetary benefits for farmers as owners of the potential urban land, this has not happened for the majority of peasants whose land was expropriated in the last four decades, perhaps except until very recently (Zhan 2017). In the name of the "public interest," the state has forcibly expropriated land from rural collectives for urban use, and it had only to compensate the affected peasants at the value of farmland, not its much higher market value as potential urban land. Farmers, even though they own their land through the collectives, actually are not allowed to dispense land as they wish, and cannot refuse state land expropriation. Stripping farmers of *de facto* control of the land, collectivization in the 1950s, relegated farmers' land ownership rights to only nominal ones (Yu 2002; Li 2010). And despite rural decollectivization in the reform era, collective ownership of rural land remains. The central government has reiterated that this ownership is fundamental and will not change for years to come (State Council 2016). This has continued to give the state considerable control over rural land as the great majority of village leaders (who hold great powers in disposing collective land) are under the Party's direct and indirect control.[15] This channel of extraction of economic resources through land transfer, from the countryside to the city, has been well maintained by the state in the reform era (Li 2010). Some of the rural residents displaced are provided with urban apartments and urban *hukou* of some sorts as compensation, but often with limited means to succeed economically in the city (Zhan 2017).

Under the dual system, the lion's share of the windfall profit through land conversion has been captured by the state, mainly local governments. In the two decades since the 1990s, the revenues collected through expropriation of rural land (and subsequent leases of the converted land) have been the most important single revenue source for many local city governments, creating a local finance system heavily dependent on land revenue (Tao et al. 2010; Tsui 2011). Furthermore, as a means to counter the economic downturn caused by the global financial crisis in 2008–09, China has greatly relaxed local governments' debt-financing, much of which is also leveraged on urban land (Liu

2018). These funds have sustained local governments' massive waves of urban construction in the last two decades. Inevitably, these forcible land takings have triggered tens of thousands of protests and resistance from peasants across the country every year (Whiting 2011; Zhan 2017).

Mechanisms and features of the dual system: a synthesis

What are the major mechanisms of the dual system through which the farm sector and its population are exploited? Below we offer a synthesis and an analysis of differences between the dual system and the Lewis model so that the functions and nature of the various processes are accurately laid bare.

Mechanisms of the dual system

Table 2, based on the work of Buckingham and Chan (2018), outlines the combination of economic and administrative means used to accomplish extraction in different eras. Labor, land (farmland and housing plots) and the agricultural product (mainly grains) are the three exploitable resources of value peasants possess. The principal mechanism is an unequal exchange of these resources between rural and urban sectors. In Mao's era, the process was a coercive extraction as it was not truly an "exchange" when the market was shut down. Peasants worked in collectives, which were required to hand to the state a certain amount of their grain harvest at state-dictated low prices.[16] In

Table 2. Mechanisms of exploiting the rural sector in different periods under the dual system.

	Mechanisms		
Target of Extraction	Unequal Sectoral Exchange	Preventing Exit from the Rural Sector or Nonagricultural "Conversions" (*nongzhuangfei*)	Monitoring and Compliance
Agricultural output (mainly in 1953–1983)	Cheap Grain: Compulsory sales of grains to the state through "price scissors". Cheap Labor: Rural labor in plentiful supply	*Hukou* system bans rural labor outflows; non-agricultural or sideline activities prohibited	Rural collectivization put the villages within the control of the state.
Rural Labor (1980-present)	Cheap Labor: Rural migrant labor in plentiful supply	Migration allowed, but the *hukou* system prevents rural laborers from *permanent* settlement in cities. Costs of social reproduction of labor are mainly outsourced to the countryside and migrant family.	City police, factory managers, employment bureaus
Farmland and rural housing plots (1992- present)	Cheap Land: State expropriation of rural land	Land law gives state the monopoly right to expropriate rural land and reap windfall profits through conversion of rural land for urban use.	Party-controlled village leaders; the court system.

Source: adapted from Buckingham and Chan (2018).

order to enforce the unequal exchange (also known as the "price scissors"), preventing exit from the system of extraction was crucial (Chen 1982; Knight and Song 1999). The measures used to accomplish this included collectivization, which not only created institutions that, for the first time in Chinese history, extended the state's reach and policing down to villages and farm households and to almost every corner of people's life, especially during the Great Leap Forward (Gao 2014), but also transferred the ownership of farmland from the family to the party-controlled collective, as explained before. The core mechanism has been to prevent rural labor and rural land from "exiting" the rural sector, or *nongzhuanfei* ("conversions to non-agricultural").[17] This mechanism effectively circumscribed the rural-*hukou* population's economic, social, and political opportunities and rights. The rural workers were then turned into a massive pool of super-low-cost labor tied to the land to produce food grains and other farm products (such as cotton for industry) for the state. In Mao's era, rural land was also limited to almost one mono-use for agriculture, the prices of which were deliberately kept low. Of course, monitoring mechanisms, such as the *hukou* registration, were established simultaneously to ensure compliance, especially to prevent rural population outflows.

In the reform era, China has embraced a partial market economy. As China has rapidly industrialized, taking advantage of world's appetite for low-priced manufactures, the dual system no longer extracts low-priced grains from the peasantry. Instead, the *hukou* system has been reformatted to become a tool for the state to extract cheap labor from a huge army of rural-*hukou* migrant labor for the urban sector, by depressing the direct and indirect costs of the workers' labor. At the same time, especially since the early 1990s, the state (mainly local governments) has also siphoned off massive revenues through expropriation of rural land, as explained earlier. The core tool of the dual system – to prevent the exit of the rural (labor and land) – is still used, but in a different manner.

The rural population is allowed to exit farming, but they cannot abandon their second-class, rural-*hukou* status. Rural-*hukou* migrants cannot formally resettle in major cities though they are part of the regular urban workforce. This prohibition serves to greatly reduce the indirect costs of their labor to urban employers and governments. Rural land is not allowed to enter the market to benefit from "*nongzhuanfei*" in an era of urbanization. Instead, more often than not, rural land is forcibly expropriated by the state for windfall profits. Essentially, the core function of the dual divide – to depress the wages of rual-*hukou* (migrant) labor and the compensation costs of rural land – has not been changed.

Social stratification and rural-urban disparities

The rural-urban divide created by the *hukou* system under the dual system is also reflected clearly in the social stratification, as illustrated in Figure 1. In Mao's era, the physical rural-urban boundary basically coincided with the rural-urban *hukou* division as the divide was enforced by bans on rural-to-urban geographical mobility. Therefore, the geographical rural-urban divide was the same as the institutional rural-urban divide along the *hukou* line. In the reform era, rural-*hukou* people are allowed to go to the city to find work, but they retain their rural-*hukou* status and are still treated as rural residents with very limited access to urban-type social security, welfare and many other public services. The physical boundary between rural and urban areas has become largely permeable, but the rural-urban divide based on the *hukou* system has not. In other words, no matter where migrants move, their *hukou* status remains the same. There is no change in the *hukou* divide. The "invisible walls" remain and effectively shut off social mobility for rural-*hukou* migrants (Chan 1994; Treiman 2012; Wu and Zheng 2016). Even more restricting to their social mobility is that the *hukou* status is hereditary by law so intergenerational social mobility is also greatly hampered (Chan and Buckingham 2008).

Regrettably, these major institutional barriers to social mobility faced by the rural-*hukou* population, some of which are insurmountable, are often missing in many narratives of China's growing middle class such as MGI (2006) and Wang (2010). In many other developing countries, rural migrants arriving in

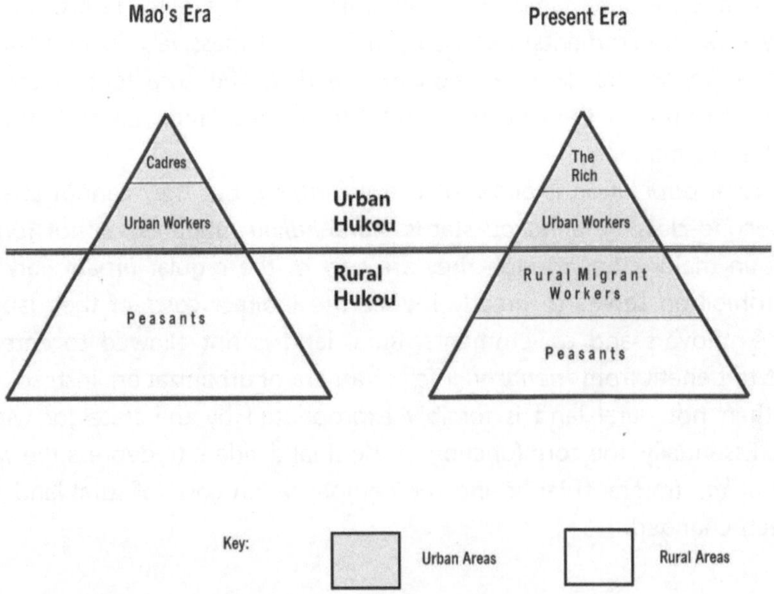

Figure 1. Social stratification and *Hukou* Divide.
Source: Chan (2009).

cities is traditionally a major source of the new middle class, as described in Saunders' (2012) uplifting accounts in *Arrival City*. In a successful, not uncommon experience in a developing country, migrants leave the farm for work in the city; they then receive a higher wage and can afford to consume more. Trading their farm jobs and subsistence lives for industrial or service jobs in the city, some migrants can then gradually live in apartments furnished with appliances and maybe even dream about sending their kids to college. This popular narrative, however, cannot be automatically applied to China because of the existence of a special rural-urban institutional divide. A number of analysts, having ignored the fundamental Chinese specificity, assume that China's rural-urban difference is simply a temporary, Lewisian divide and that a large portion of the huge numbers of rural migrants arriving in the city will soon become the middle class or the urban consumer class.[18]

Given such social stratification and state policy in buttressing the rural-urban divide, it is not surprising that China has experienced an unusually high level of rural-urban inequality and wide socioeconomic cleavage, even by developing countries' standards, as has been widely documented (e.g. Knight and Song 1999; Yang and Cai 2003; Whyte 2010b; Logan 2015). The disparities are so large that Treiman (2012) aptly calls them "between heaven and earth." This rural-urban division is also clearly reflected in the *hukou*-based disparities in education at all levels, in health outcomes, and in highly segmented *hukou*-based marriage markets (Lu 2004; Whyte 2010a; Wang and Schwartz 2015; Lui 2016). Indeed, even as recently as 2017, Fei-ling Wang, a critic of the *hukou* system, has these strong words to say about the situation: "The urban-rural difference in China is horrendous – living standard, income, education, medical service, you name it." (Juneja 2017).

The Lewis model and China's rural-urban dual system

The term rural-urban "dualistic sector" or "dualism" – essentially based on the Lewis (1954) model of unlimited rural surplus labor – has been widely used by scholars both within China and outside to explain China's rural-urban labor mobility and, indeed, has gained orthodoxy. The model, based on neoclassical economics and named after Nobel laureate in economics Arthur Lewis, posits a rural-urban transition based on two sectors in the market economy: the modern industrial sector and the traditional subsistence agricultural sector. There is an initial large, "unlimited" rural labor surplus in the agricultural sector, typical of many developing countries. The model explains the advantage of the industrial sector (industrial capitalists) over the agricultural sector on the basis of the supply and demand for rural surplus labor. In his model, the dualistic structure is a natural outcome of the initial differential levels of technological development between the industrial and the agricultural sectors in the early stages of industrialization (Todaro 1977). The rural-urban differences, considered as the initial outcome of the agricultural-industrial transition, will disappear when the rural labor surplus is exhausted, leading to rural-urban integration.[19] The Lewis model, postulating several major features similar to those of China's dual economy, definitely has heuristic value for understanding China (see e.g. Cai 2007; Chan 2010b; Kroeber 2010). There are, however, fundamental differences between the Lewis dual-sector model, or broadly, "dualism" in the development literature, and China's dual system (Yu 1994; Deng 2012; Qi and Li 2019), pointing to a rather dissimilar nature of the Chinese practice.

(i) The Chinese dual system does not just simply feature a rural-urban divide in the economy; the dualism is far more encompassing, covering not only the economy but also a very large part of society, extending indeed to the political, social and cultural (Bai 2012). The Lewis model, however, is entirely about the labor supply for the urban sector driven by market forces, on the basis of "free choice of individuals" (Arrighi 1970).

(ii) More importantly, the Chinese dual system is part and parcel of a coercive state strategy to industrialize the country, quite different from the market-driven Lewis model, in which the rural-urban divide will collapse once the rural labor surplus is exhausted. The Chinese dual system is designed to compel the peasantry into making "tribute"; the mechanisms used are despotic, as in forced collectivization. Even in the reform era when there is a partial market, there are still major differences, as explained below.

(iii) The Lewis model is premised on a single "society": rural and urban residents, who, though economically different and operating separately (initially) in two sectors, are treated equally by law and compete freely in the market. In short, it is "one country, one system" despite the rural-urban differences. The Chinese dual system, on the other hand, created two separate "societies" and reinforced the divide throughout the entire PRC period (Shen 1997; Whyte 2010b). Rural-*hukou* and urban-*hukou* people are treated differently by law, forming "one country, two systems" (Chan 1999). Indeed, Chinese scholars talk about the dual system as a dual "social system" (*shehui tizhi*), one which is enforced also politically and socially, far broader than Lewis' dual economy (Shen 1997; Guo 2017). The goal of the dual system was to create a disenfranchised class to be exploited. A rural *hukou* is also part of a system of the ascribed *shenfenzhi* (identity status). The starkly different fates and opportunities of the two populations are first determined by the state, as shown in Table 1 and Figure 1. Even if the opportunities to out-migrate are available in the reform era, the rural-*hukou* migrants remain a second-class population in the city, analogous to Africans under the apartheid in South Africa or African-Americans in the Jim Crow era in the USA (Alexander and Chan 2004; Chan 2017).

(iv) The Lewis model is a model of transition (to integration); China's dual system is a system of "division and exclusion," not integration, to use Wang's (2005) words. The former postulates that the dualism is only temporary. As industrialization and urbanization proceed, economic dualism will end eventually. Its "one system" also rests on the assumption of social integration following economic integration as subsistence farmers turn into urban wage workers. In the Lewisian transition, the universalization of (urban) wage labor is usually directly linked to the extension of the citizenship rights (Friedman 2017). In several countries in East Asia, this rural-urban transition takes about two to three decades to complete (see e.g. Wantanabe 1994). In post-1949 China, the rural-urban cleavage is an outcome of the willful acts of the state; the rural-urban disparities are systematic and chronic, not transitional. Indeed, the dual structure has been pushed further to become a "bi-polar structure" (Yu 1994), further widening the rural-urban gap.

Therefore, it is not surprising that the dual system and the accompanying economic and social dualism have existed more than 60 years, despite the appearance of the rural-*hukou* labor shortages in the age group below 30 in the early years of the 21st century (Cai 2018). The dual system is likely to continue for at least another couple of decades, even in the most optimistic prognosis. The longevity speaks to the fundamental differences between the Lewis model and Chinese case. By 2016, about 170 million rural-*hukou* workers have become mostly urban wage labor but there is still very limited social integration (Chan and Yang 2019; Wu 2017). Just like in the USSR, the dualism in China emerged and has persisted not simply as a "natural" (market) condition, but mainly as a product of the state's deliberate strategy. Indeed, Chan and Buckingham (2008), Whyte (2010a) and Chan (2010b) have shown that several of the recent state policies have continued to reinforce and prolong the dualism.

(v) In the present era, the market has a mixed effect on rural-urban equalization: on the one hand, the labor mobility has helped to reduce the rural-urban wage gap; on the other, under the condition of rural labor surplus for most of the time since 1979, rural labor mobility has greatly benefited capital, as explained in the Lewis model. Oftentimes, the dual system creates the primary platform upon which capitalists and urban governments are able to exploit rural-*hukou* labor in the city. A good example of this is the low status of the rural migrant workers. Many have argued that their predicament is caused mainly by the market or global capital (e.g. Harvey 2005; Pun and Chan 2012). This is true to some extent, but a deeper analysis shows that it is the state – through the dual system established under Mao and its various derivative institutions and policies – which created a much more embracing administrative, economic and legal structure through which capitalists have exploited migrants

(Lee and Kofman 2012). The reason why the rural-*hukou* population has become a *permanent* second class so easily exploitable is because state institutions have disfranchised the rural-*hukou* population in the first place and continues to buttress that asymmetry, enabling that population to be the prey of capital and the market.

(vi) The special nature of China's dual system has led several Chinese scholars, such as Fan and Zhao (1988), Wang (2016), and Ren (2017) to call the system a "double dual system" (*shuangchong eryuanzhi*), referring to both the economic dualism common in the development process and the much larger dualism set up by the state, as part of the socialist development strategy. Based on that, one can characterize China's rural-urban dual system as a *mega* dual system, different from the more limited, "smaller" dualistic structure of Lewis. The latter is driven totally by neoclassical economics while the former is a product of social relations and political structure (Qi and Li. 2019). This means that the rural-*hukou* population suffers doubly from both state-based *and* market-based exploitation and oppression.

Table 3 summarizes the main differences of the two models. The various points can also be illustrated by the plight of the children of the rural-*hukou* migrants in their growing up, especially in relation to education. It is estimated that the population size of children of China's current 170 million migrant laborers are about 100 million, or one-third of China's children population (Chan 2018a). These are children whose lives have been profoundly disrupted, mostly negatively by the split families as their parents migrate to work in urban areas under China's current system. These children have become victims of the urbanization process under the dual system, in addition to suffering market-based disadvantages, as in any other rural-urban migration in the development process (Ye et al 2013). The two major groups of children of

Table 3. Comparison of the Lewis model and China's rural-urban dual system.

	The Lewis Model	China's Rural-urban Dual System
1. Coverage	The economy only (based on neoclassical economics)	The political economy and society (based on power structure, social relations and economics)
2. Drivers	Industrial growth through the market (supply and demand of labor) based on the "free choice of individuals"	Industrial growth based on coercion through administrative means, as part of the state strategy
3. Social structure	One society of equal rights, etc	Two different societies of different rights, etc.
4. Labor mobility	Free	Banned under Mao; highly regulated after Mao. Peasants cannot exit from the rural-*hukou* status.
5. Process and temporality	Temporary transition to integration	Permanent division and exclusion

migrants are the "migrant children" and the "left-behind children" (LBC), of sizes of 34 million and 69 million, respectively, based on the 2015 mini-census (Chan and Ren 2018).

To escape poverty, most of the rural-*hukou* population have to find work in the city to generate the needed income for their family. In the last two decades, rural-urban migration in China has increasingly involved married couples and their children as the migrant labor population matures and migrants begin to settle, i.e. increasing "familization of migration" (Qu, Yang, and Cheng 2015; Duan Chenggrong, quoted in Wang 2016). However, unlike in other countries, most migrant workers moving to cities to work cannot use urban public services, including public education for their children. As expected, children accompanying migrant parents face many educational obstacles in the destination. Until the late 2000s, it was very hard for migrant children to be enrolled in public schools in cities, and a great majority of them could only go to make-shift "private" migrants' schools. In the late 2000s, public schools in some cities began to accept migrant children, following a central policy issued in 2008 (Ming 2013; Yang 2016). But the enrollment is limited to only grades 1–9, with a high admission eligibility bar. Consequently, many children of migrant workers are simply kept out of schools or have to go back to their home village to continue their education, especially after grade 9. Even for those luckier ones admitted to urban schools, they often face acute discrimination because of the segregated classrooms and playgrounds in public schools, or they are simply sequestered in migrant children-only schools in the outlying areas of the cities (Lan 2014).

As a result, in addition to the "usual" obstacles faced by new arrivals in the city from the countryside and being poor, children of Chinese migrant labor face huge, often insurmountable, institutional (and legal) barriers in getting an education in the city. This, combined with the "market" obstacles (such as not being able to afford reasonable family housing,[20] and long working hours that leave parents with no time for caring their children), have forced migrant workers to leave their children behind in the countryside. To no one's surprise, LBC face many serious problems growing up in the countryside in *de facto* single-parent or no-parent families (Normile 2017).

The demography in Shanghai in 2010 offers a glimpse of this "purging" of children of the migrant workers in the city. An estimation based on the 2010 Census data suggests that the 8.85 million adult migrant population in Shanghai would have yielded a children's population (those under age 18) of 3.53 million (based on the then migrants' fertility rates) (Chan 2018a). But only 0.96 million were actually present in Shanghai in 2010, implying that 2.57 million were left in their parents' *hukou* home locations (villages) or elsewhere. Not only are migrant workers exploited in the city under that system, but their children are also poorly educated and grow up in very undesirable environments whether in the countryside or in the city. Because of the precarious nature of their parents' work tenure, many children of migrants also lead an unstable life, shifting between being "migrant children" and LBC, and from place to place. Indeed, the two groups are often interchangeable, and based on figures in Chan and Ren (2018), it can be estimated total LBC population, those who stayed with one or no parents at all on a daily basis, surged from about 37 million in 2000 to 88 million in 2015, quite contrary to the official narratives of a declining LBC population. That also means about 85 percent of migrant children live in *de facto* single-parent or no-parent families! It is not hard to imagine that most of these children will likely become the *de facto* army of reserve labor in the future, repeating the destiny of their parents (Xiong 2015; Pan 2018). The impact of recent Chinese education measures and the *hukou* measures is to reinforce a pattern of rural-*hukou* children only finishing lower middle school, but urban children increasingly getting college education, with the lion's share of the benefit of the university expansion in the last two decades going to the urban *hukou* population.

This is the power of the dual system, and the oppression faced by the rural-*hukou* population is far more than that in the typical market or capitalist system. Our analysis also refutes the argument that rural migrant workers leave their children behind in the countryside as a "wise" household strategy to maximize their work incomes while also simultaneously taking advantage of low-cost rural life, forwarded by, for example, Schmidt-Kallert and Franke (2013), Zhu and Li (2014), and Chen and Fan (2016). This narrow "economic calculus" interpretation ignores the structural oppression of the rural-*hukou* population, including many economic, social and legal constraints the group faces under the dual system (Yang 2015; Ye, Wu and Meng 2015; Wei and Gong 2019). Evidently, the thesis based on a micro economic rationality fails to recognize the power structure and the highly exploitative and unequal nature of China's mega development model.

Conclusion

Based on the work of previous scholars, this paper has taken a broader, historical perspective and analyzed the roots and nature of the Chinese rural-urban dual system, and its Leninist-socialist development logic, operations and formats in different eras since 1949. The system has its genealogy dating back to Stalin's industrialization in the USSR in the late 1920s, which China followed in the 1950s. The system has erected a wall between the rural and urban sectors to produce a highly exploitable rural "society" and population. The major purpose is to enable the state and the urban-industrial sector to siphon off resources from the rural sector. One of the major hallmarks of this dual system in the USSR under Stalin and China under Mao was to restrict the geographical mobility of rural residents and tie most of them to production in the countryside (Edelman 2013). The system has played a crucial role not only in the industrialization push under Mao in China, but also in its rapid economic rise since the 1980s. In the last forty years, the rural-urban divide mechanisms have metamorphosed to suit the new historical and geographical circumstances to produce massive low-wage migrant labor for the global economy and to capitalize rural land to generate revenues for the new urbanization quest. The dual system is robust and adapts to the new hybrid economy so as to continue serving the socialist state agenda.

The Chinese dual system is neither simply a Lewisian rural-urban divide in a market-driven dual-sector economy nor a product of China's neoliberalist shift. Instead, the Chinese system is part and parcel of a Stalinist-socialist state strategy to industrialize the country by dividing the country into "two systems" to create a second-class population so that their labor, property and products are easily subject to exploitation or expropriation. That despotic system is also far more encompassing, covering not only the economy but also politics and the society. As such, it has led to the oppression of the rural sector and population on a scale far exceeding what one would usually find under a capitalist, market system. In the present era of urbanization, the dual system has enabled the state to outsource almost the full costs of social production of migrant labor to the countryside, as well as to expropriate rural land for development at low costs.

Our paper has also demonstrated the tenacious historical continuity and pliability of the socialist dual system and one of its major components, the *hukou* system, in shaping China's development and rural and urban landscapes in the last six decades. The present dual system is not just an institutional hangover of Maoism, awaiting the power of the market to sweep its remnants into history. Instead, the dual system as a Leninist-socialist tool of "development" is alive and still strong today, though in a different form. Our research has challenged a large urban studies and geographical literature resting on the paradigms of market transition, neoliberalism, and migrant assimilation. Without reference to the history of the planned economy under socialism, those studies assume that China is a capitalist country "with Chinese characteristics," or at least is moving on the path of transition to market capitalism and "integrating" migrants into the cities. As with Chan (2009), and Buckingham (2017), our analysis has shown that those theorizations to squeeze the Chinese case into the frameworks of the market transition or global neoliberalism are highly problematic. The primacy of the dual system remains central in explaining the plight of the rural-*hukou* population. In short, the "two systems" still remain strong; it is not only too early, but also quite wrong, to sound the death knell of the *hukou* system, as in some popular narratives such as Xinminwang (2014) and Lau (2015) would mislead us. We hope our empirical and theoretical analyses have made clear the nature of the Chinese development model and presented a broader perspective, from which one can take the Chinese urban studies scholarship to the next level.

Notes

1. There are at least two major strands of socialism commonly agreed upon: one is Bolshevik/Leninist socialism of one party monopoly and party-state, such as the system practiced in the USSR and China; the other is socialism achieved through a competitive multi-party political system in a capitalist country, often called "socialist democracy", arguably influenced by works of Eduard Bernstein and Karl Kausky. The latter exists in Western and Northern Europe. Despite the embrace of the market in the reform era, China has repeatedly proclaimed in all major party and government documents that it is practicing "socialism with Chinese characteristics" with the one-party system and party-state established under Mao remaining intact. The Chinese government calls its economy a "socialist market economy," highlighting the supremacy of the socialist/communist party in the economy. At the second plenum of the 19th Party Congress in 2018, the Party amended the Chinese Constitution to clearly state that "(t)he leadership of the Chinese Communist Party is the most essential feature of socialism with Chinese characteristics." (*Constitution* 2018, Article 2).
2. Exploiting the peasantry in early industrialization also happened in a number of capitalist states such as Korea, Taiwan under Japan, and some Latin American countries. They were predominantly rural in the early 20th century, and their industrialization was funded by extractive rural-urban transfers (see Lipton 1977).
3. This refers to the problems faced by agriculture, villages, and the peasantry, and loosely to the problems faced by the rural sector in general. See Ye (2018) for a discussion.
4. Others include: Yu (1994), Wang and Wang (1996), Liu (1996), Shen (1997), and SCDRC (2014).
5. For example, the 3rd Plenum of the 18th Party Congress held in 2013 stated clearly that the rural-urban dual structure remained a major obstacle to rural-urban integration (CCP 2013; Chan 2013). The 16th Party Congress Report in 2002 also stated that "the urban-rural dual structure had not been changed" (quoted in SCDRC 2014, p.7).
6. Critiques of pre-1990 Western scholarship on China's urbanization and development are found elsewhere, such as (Chan 1994). Other reviews of specific problems in recent scholarship are: on development paradigms and rural/urban *hukou*-based population statistics in Chan (2007), Whyte (2010a), and Gibson and Li (2017); on errors in interpreting China's recent *hukou* reforms in Chan and Buckingham (2008). For a more general critique, see Palmer (2015). Following Western mainstream literature, much of the geography scholarship coming out of China also suffers from similar problems with notable exceptions such as Wang and Liu (2015).
7. Examples abound, from the more theoretically-oriented such as Harvey (2005), Pun (2005), Wu (2008), He and Wu. (2009), and Webber (2008), to more empirical works such as Campanella (2008), Wang (2010), Webster (2014) and Wissink, Hazelzet, and Breitung (2014).
8. The problems with the use of the term *chengzhongcun* are examined by: Chung (2010), and Buckingham and Chan (2018); on statistics, by Chan and Xu (1985), Chan (2007), and Gibson and Li (2017).

9. Indeed, during its formative years in the 1920s and 1930s, the CCP received close guidance, directives and support from the USSR. The first communist base (government) set up by Mao in Jiangxi in 1931–37 was named "Chinese Soviet Republic."
10. The Soviet rapid industrialization generated extensive recruitment of rural labor to work in factories, mines, etc. so that large numbers of peasants were transferred to the urban sector in the 1930s and after WWII. There was a somewhat similar wave of recruitment of farmers to work in mines and factories in the 1950s in China. These farmers became industrial workers and also got their urban *hukou*. However, the demography of the two countries was quite different, with USSR having a smaller population and facing a much lower population pressure on land than China, the USSR did not implement the migration control as rigidly and stringently as in China in the 1960s and 1970s.
11. Tiebout (1956) posits that people, facing unfavorable local state actions (especially imposition of high taxes and inadequate provisions of public services) can move to another locale where taxes and public services are better suited to their preferences when the option of migration exists. This action is often likened to "voting with their feet" as if someone exercising his voting right via "exit."
12. More precisely, the differentiation was between the "agricultural" and "non-agricultural" *hukou*. For details, see Chan and Buckingham (2008).
13. But even as of today, China is far from being a market economy with the state playing a pivotal role (Lardy 2019).
14. They call this process part of the "capitalist transition" and argue that it is a fundamentally different the process in Mao's era. We disagree with that characterization (see Section 5). A full treatment is beyond the scope of this paper.
15. Inevitably, there are local (village) leaders acting on their own, some to the dismay of their supervisory party/governments. These local leaders are often punished when their deviance is discovered. See more about the "game play" (*boyi*) among different levels of local officials and leaders in Gao and Su (2019).
16. Collectives were residual claimants on the harvest after fulfilling the state procurement targets. High extraction was a major reason for the catastrophic famines in 1959–61, causing 25–35 million deaths (Yang 2012; Dikötter 2010).
17. Detailed explanations of the *nongzhuanfei* process applied to rural labor are in Chan and Zhang (1999), and Chan and Buckingham (2008); and to land, in Li (2005).
18. For example, Brookings analysts Kharas and Gertz (2010) put forward a rosy forecast and argued that China would add 500 million people to its middle class between 2010 and 2020 simply because of the rapid rural-urban migration. Stephen Roach (2012), a financial guru, writing in 2012, also contended that China's large number of rural-urban migrants would quickly help China solve the "ghost towns" problem, soaking up the excess housing supply, and turn them into "thriving metropolitan areas". See a critique of their work in Chan (2012).
19. This point when integration begins is the "Lewis turning point." According to Lewis, industrial wages in a developing country starts to rise at that point when the supply of surplus labor from the rural areas tapers off. Chan (2010b) argues that the *hukou* system has extended the period before reaching the turning point in China.
20. Note that rural-*hukou* migrants are not eligible for subsidized social housing despite being poor. Most of them live in dormitories or are cramped in very crowded, often informal, housing.

Acknowledgments

Earlier versions of this paper has been presented at Hong Kong Baptist University, Taiwan National University, and University of Washington. We appreciate the comments and suggestions made by the audiences. Carolyn Cartier, Eli Friedman, James Lin, Tim Oakes, Dorothy Solinger, Wing-shing Tang, and Martin Whyte read an earlier version and offered useful comments, for which we are most grateful.

Disclosure statement

No potential conflict of interest was reported by the authors.

Funding

This work was supported by a China Studies faculty research grant provided by the Jackson School of International Studies, University of Washington.

References

Alexander, Peter, and Anita Chan. 2004. "Does China Have an Apartheid Pass System?" *Journal of Ethnic and Migration Studies* 30 (4): 609–629. doi:10.1080/1369183041000169948.

Allen, Robert. 2009. *Farm to Factory: A Reinterpretation of the Soviet Industrial Revolution*. Princeton, NJ: Princeton University Press.

Arrighi, Giovanni. 1970. "Labour Supplies in Historical Perspective: A Study of Proletarianization of African Peasantry in Rhodesia." *Journal of Development Studies* 6 (3): 189–234. doi:10.1080/00220387008421322.

Arrighi, Giovanni. 2007. *Adam Smith in Beijing: Lineages of the Twenty-first Century*. London: Verso.

Bai, Yongxiu. 2012. "Chengxiang Eryuan Jiegou De Zhongguo Shijiao: Xiangcheng、Tuozhang、Lujing." [A China's Perspective of the Urban-rural Dual Structure: Formation, Extension and Path.] *Xueshu Yuekan [Academic Monthy]* 44(5): 67–76.

Bernstein, Thomas. 2014. "Introduction: The Complexities of Learning from the Soviet Union." In *China Learns from the Soviet Union, 1949–Present*, edited by T. Bernstein and H.-Y. Li, 1–15. London: Rowman and Littlefield.

Bideleux, Robert. 1985. *Communism and Development*. London: Methuen.

Bo, Yibo. 1991. *Ruogan Zhongda Juece Yu Shijian De Huigu* [A Memoir of Certain Major Historic Decisions and Events]. Beijing: History of Chinese Communist Party Press.

Boettke, Peter. 1990. *The Political Economy of Soviet Socialism: The Formative Years, 1918-1928*. New York, NY: Springer.

Brown, Jeremy. 2012. *City versus Countryside in Mao's China: Negotiating the Divide*. New York, NY: Cambridge University Press.

Buckingham, Will. 2017. "Uncorking the Neoliberal Bottle: Neoliberal Critique and Urban Change in China." *Eurasian Geography and Economics* 58 (3): 297–315. doi:10.1080/15387216.2017.1350196.

Buckingham, Will, and Kam Wing. Chan. 2018. "One City, Two Systems: *Chengzhongcun in* China's Urban System." *Journal of Contemporary China* 27 (2): 584–595. doi:10.1080/10670564.2018.1433576.

Bukharin, Nikolai. 2007. *The Politics and Economics of the Transition Period*. New York, NY: Routledge.

Burdekin, Richard. C. K. 1989. "Preobrazhensky's Theory of Primitive Socialist Accumulation." *Journal of Contemporary Asia* 19: 297–307. doi:10.1080/00472338980000341.

Cai, Fang. 2007. "Zhongguo Jingji Fazhan De Liuyisi Zhuanzhedian [China's Economic Development and the Lewis Transition Point]." In *Zhongguo Renkou Yu Laodong Lupishupishu*[Green Book of Population and Labor] edited by F. Cai, 147–169. Vol. 8. Beijing, China: Shehui Kexue Wenxian Chubanshe.

Cai, Fang. 2018. "Zhongguo Gaige Kaifang Sishi Buhuo" [China's Forty Years of Reform and Opening]." In *40 Years' Reform and Development of China's Economy (1978-2018)*, edited by F. Cai, 1–19. Beijing: Shehui Kexue Wenxian Chubanshe.

Campanella, Thomas. J. 2008. *The Concrete Dragon: China's Urban Revolution and What It Means for the World*. New York: Princeton Architectural Press.

Cartier, Carolyn. 2015. "Territorial Urbanization and the Party-State in China." *Territory, Politics and Governance* 3 (3): 294–320. doi:10.1080/21622671.2015.1005125.

Cartier, Carolyn. 2019. "Exemplary Cities in China: The Capitalist Aesthetic and the Loss of Space." *Eurasian Geography and Economics* 60 (4).

CCP (Chinese Communist Party). 2013. "The Decision on Major Issues Concerning Comprehensively Deepening Reforms." 16 November. Accessed 2 December 2013. http://www.china.org.cn/china/third_plenary_session/2013-11/16/content_30620736.htm

Chan, Anita, and Robert J. S. Ross. 2003. "Racing to the Bottom: Industrial Trade without a Social Clause." *Third World Quarterly* 24 (6): 1011–1028. doi:10.1080/01436590310001630044.

Chan, Kam Wing. 1994. *Cities with Invisible Walls: Reinterpreting Urbanization in Post-1949 China*. Hong Kong: Oxford University Press.

Chan, Kam Wing. 1999. "'One Country, Two Systems': Rural-Urban Dualism in the PRC." In *PRC Political Economy: Prospects under Jiang Zemin*, edited by H.-H. Wu., 169–189. Taiwan: Mainland Affairs Council.

Chan, Kam Wing. 2007. "Misconceptions and Complexities in the Study of China's Cities: Definitions, Statistics, and Implications." *Eurasian Geography and Economics* 48 (4): 383–412. doi:10.2747/1538-7216.48.4.383.

Chan, Kam Wing. 2009. "The Chinese *Hukou* System at 50." *Eurasian Geography and Economics* 50 (2): 197–221. doi:10.2747/1539-7216.50.2.197.

Chan, Kam Wing. 2010b. "A China Paradox: Migrant Labor Shortage Amidst Rural Labor Supply Abundance." *Eurasian Geography and Economics* 51 (4): 513–530. doi:10.2747/1539-7216.51.4.513.

Chan, Kam Wing. 2010b. "Fundamentals of China's Urbanization and Policy." *The China Review* 10 (2): 63–94.

Chan, Kam Wing. 2010c. "The Global Financial Crisis and Migrant Workers in China: There Is No Future as a Labourer; Returning to the Village Has No Meaning." *International Journal of Urban and Regional Research* 34 (3): 659–677. doi:10.1111/j.1468-2427.2010.00987.x.

Chan, Kam Wing. 2011. "In the City, but Not of the City: The Myth of China's Urbanization." Accessed 29 August 2017. http://www.chinausfocus.com/political-social-development/in-the-city-but-not-of-the-city-the-myth-of-china%E2%80%99s-urbanization/

Chan, Kam Wing. 2012. "Crossing the 50 Percent Population Rubicon: Can China Urbanize to Prosperity?" *Eurasian Geography and Economics* 53 (1): 63–86. doi:10.2747/1539-7216.53.1.63.

Chan, Kam Wing. 2013. "Zhongguo Chengren Chengxiang Eryuan Jiegou De Zhiyue Shi Juda De Jinbu [China's Recognition of the Constraint of the Rural-urban Dual Structure is a Huge Improvement]." *Renminwang*, 13 November. Accessed 4 November 2013. http://politics.people.com.cn/n/2013/1113/c1001-23525599.html

Chan, Kam Wing. 2017. "Chi 'Danchengpiao' De Qianxizhe: Laolunsi 'Qianyi Xilie' De Qishi ["Migrants with "One-way Tickets": What Jacob Lawrence's *Migration Series* Has

Revealed]." *Caixin*, Online 13, April 2017. Accessed, http://opinion.caixin.com/2017-04-13/101078036.html

Chan, Kam Wing. 2018a. "Introduction: China's *Hukou* System, Migration and Urbanization." In *Urbanization with Chinese Characteristics: The Hukou System and Migration*, edited by Kam Wing Chan, 1–18. Oxon, UK: Taylor & Francis.

Chan, Kam Wing. 2018b. *Urbanization with Chinese Characteristics: The Hukou System and Migration*. Oxon, UK: Taylor & Francis.

Chan, Kam Wing, and Xiaxia Yang. 2019. "Internal Migration in China: Trends, Geography and Major Issues." In *Routledge Handbook of Migration and Development*, edited by T. Bastia and R. Skeldon. Routledge (in press).

Chan, Kam Wing, and Li Zhang. 1999. "The *Hukou* System and Rural Urban Migration: Processes and Changes." *The China Quarterly* 160: 818–855. doi:10.1017/S0305741000001351.

Chan, Kam Wing, and Will Buckingham. 2008. "Is China Abolishing the *Hukou* System?" *The China Quarterly* 195: 582–606. doi:10.1017/S0305741008000787.

Chan, Kam Wing, and Xueqiang Xu. 1985. "Urban Population Growth and Urbanization in China since 1949: Reconstructing a Baseline." *The China Quarterly* 104: 583–613. doi:10.1017/S0305741000033324.

Chan, Kam Wing, and Yuan Ren. 2018. "Children of Migrants in China in the 21st Century: Trends, Living Arrangements, Age-gender Structure, and Geography." *Eurasian Geography and Economics* 59 (2): 133–163. doi:10.1080/15387216.2018.1535906.

Chao, Kang. 1986. *Man and Land in Chinese History*. Stanford, California: Stanford University Press.

Chen, Chuanbao, and C. Cindy Fan. 2016. "China's Hukou Paradox: Why Don't Rural Migrants Want Urban Hukou?" *The China Review* 16 (3): 9–39.

Chen, Kaiguo. 1982. "A Tentative Inquiry into the Scissors Gap in the Rate of Exchange between Industrial and Agricultural Products." *Social Sciences in China* 3 (2): 55–74.

Chuang, Julia. 2015. "Urbanization through Dispossession: Survival and Stratification in China's New Township." *The Journal of Peasant Studies* 42 (2): 275–294. doi:10.1080/03066150.2014.990446.

Chung, Him. 2010. "Building an Image of Villages-in-the-City: A Clarification of China's Distinct Urban Spaces." *International Journal of Urban and Regional Research* 34 (2): 421–437.

Conquest, Robert. 1987. *The Harvest of Sorrow: Soviet Collectivization and the Terror-famine*. New York: Oxford University Press.

Constitution, the People's Republic of China. 2018. Accessed 15 March 2018. http://www.xinhuanet.com/politics/2018lh/2018-03/22/c_1122572202.htm

Deng, Jie. 2012. "Insiders' Entitlements: Formation of the Household Registration (*Huji/hukou*) System in Shanghai, People's Republic of China (1949-1959)." Ph.D Dissertation in History, Queen's University, Canada. doi:10.1094/PDIS-11-11-0999-PDN.

Dikötter, Frank. 2010. *Mao's Great Famine: The History Of China's Most Devastating Catastrophe, 1958-1962*. New York: Walker & CO.

Edelman, Marc. 2013. "What Is A Peasant? What Are Peasantries? A Briefing Paper On Issues Of Definition." Presented at the Intergovernmental Working Group on a United Nations Declaration on the Rights of Peasants and Other People Working in Rural Areas, Geneva, Switzerland, 15-19 July 2013.

Fan, Di, and Zhao Chen. 1988. "Guanyu Shuang Eryuan Jingji Jiegou De Lilun Fenxi [Analysis on the Theory of Double Dual-Economy Structure]." *Jingji Yanjiu [Economic Studies]* 4: 61–68 & 18.

Fitzpatrick, Sheila. 1996. *Stalin's Peasants: Resistance and Survival in the Russian Village after Collectivization*. New York, NY: Oxford University Press.

Friedman, Eli. 2017. "Just-In-Time Urbanization? Managing Migration, Citizenship, and Schooling in the Chinese City." *Critical Sociology*, 44(3): 503–518. doi: 10.1177/0896920517695867.

Gao, Hua. 2014. *Dayuejin Yundong Yu Guojia Quanli De Kuozhang*. [Great Leap Forward and the Expansion of State Power]. Guo Hua. *Lishi Biji [Notes on History]*. I. Hong Kong: Oxford University Press. 333–354.

Gao, Hua. 2018. *How Did the Red Sun Rise? the Origin and Development of the Yan'an Rectification Movement, 1930-1945*. Hong Kong, China: Chinese University Press.

Gao, Jia, and Yuanyuan Su. 2019. *Social Mobilisation in Post-Industrial China: The Case of Rural China*. Cheltenham, UK: Edward Elgar.

Gardner, Bradley M. 2017. *China's Great Migration How the Poor Built a Prosperous Nation*. Oakland, California: Independent Institute.

Ge, Jianxiong, Songdi Wu, and Shuji Cao. 1997. *Zhongguo Yimin Shi*. [China Migration History]. Fuzhou: Fujian Renmin Chubanshe.

Gibson, John, and Chao Li. 2017. "The Erroneous Use of China's Population and per Capita Data: A Structured Review and Critical Test." *Journal of Economic Surveys* 31 (4): 905–922. doi:10.1111/joes.12178.

Gregory, Paul R, and Robert C. Stuart. 1986. *Soviet Economic Structure and Performance*. New York: Harper & Row Publishers.

Guo, Qiang. 2017. "Nongmingong Shiminhua Yicheng Gaige Hexin Wenti." [Citizenization is a Core Question in Reform.] *Caixin*, 15 July. Accessed 20 July 2017. http://opinion.caixin.com/2017-07-15/101116294.html?sourceEntityId=101120084

Guo, Shutian, and Chunbin Liu. 1990. *Shiheng De Zhongguo*. [An Unbalanced China]. Shijiazhuang: Hebei Renmin Chubanshe.

Harney, Alexandria. 2008. *The China Price: The True Cost of Chinese Competitive Advantage*. London: Penguin.

Harvey, David. 2005. *A Brief History of Neoliberalism*. Oxford: Oxford University Press.

He, Shenjing, and Fulong Wu. 2009. "China's Emerging Neoliberal Urbanism: Perspectives from Urban Redevelopment." *Antipode* 41: 282–304. doi:10.1111/anti.2009.41.issue-2.

Huang, Youqin, and Chengdong Yi. 2014. "Invisible Migrant Enclaves in Chinese Cities: Underground Living in Beijing, China." *Urban Studies* 52 (15): 2948–2973. doi:10.1177/0042098014564535.

Huntington, Samuel P. 1968. *Political Order in Changing Societies*. New Haven, CT: Yale University Press.

Juneja, Priyanka. 2017. "China's Hukou System: An Interview with Fei-ling Wang." *The Diplomat*, 14 July. Accessed 16 July 2017. http://thediplomat.com/2017/07/chinas-hukou-system/

Kharas, Homi, and Geoffrey Gertz. 2010. "The New Global Middle Class: A Crossover from West to East." In *China's Emerging Middle Class: Beyond Economic Transformation*, edited by C. Li, 32–51. Washington, D.C.: Brookings Institution.

Knight, John, and Lina Song. 1999. *The Rural-urban Divide: Economic Disparities and Interactions in China*. Oxford, UK: Oxford University Press.

Kornai, Janos. 1992. *The Socialist System: The Political Economy of Communism*. Princeton: Princeton University Press.

Kroeber, Arthur. 2010. "The End of Surplus Labor." *China Economic Quarterly* 1: 35–46.

Lan, Pei-chia. 2014. "Segmented Incorporation: The Second Generation of Rural Migrants in Shanghai." *The China Quarterly* 1017: 1–23.

Lardy, Nicholas. 1985. "State Intervention and Peasant Opportunities." In *Chinese Rural Development The Great Transformation*, edited by W. Parish, 33–56. Armonk: M.E. Sharpe.

Lardy, Nicholas. 2019. *The State Strikes Back: The End of Economic Reform in China?* Washington, D.C.: Peterson Institute for International Economics.

Lau, Chong-chor. 2015. "Eryuan Shehui De Zhongjie [The End of the Dual Society]." *Singtao Daily* 3 March. B3.

Lee, Ching Kwan. 1998. *Gender and the South China Miracle*. Berkeley, CA: University of California Press.

Lee, Ching Kwan, and Yelizavetta Kofman. 2012. "The Politics of Precarity: Views beyond the United States." *Work and Occupations* 39 (4): 388–408. doi:10.1177/0730888412446710.

Levine, Steven, and Gibert Rozman. 1988. "The Chinese Debate about Soviet Socialism, 1978-1985." *Political Science Quarterly* 103 (4): 760–780. doi:10.2307/2150928.

Lewis, Arthur. 1954. "Economic Development with Unlimited Supplies of Labour." *The Manchester School* 22 (2): 139–191. doi:10.1111/j.1467-9957.1954.tb00021.x.

Li, Fengzhang. 2010. "Jiti Tudi Suoyouzhi De Benzhi Jiqi Biange [The Nature of Collective Land Ownership and Evolution]." *Fazhi Yu Shehui Fazhan [Law and Social Development]* 5: 16–30.

Li, Xinan. 2005. "Chengshihua, Tudi, "nongzhuanfei" Yu Shidi Nongmin De Baozhang Wenti [Question on Urbanization, Non-agricultural Conversion of Land, and Dispossessed Peasants]." *Ningxia Shehui Kexue* 5: 67–72.

Lin, George. C. S. 2009. *Developing China: Land, Politics and Social Conditions*. Oxon, UK: Routledge.

Lin, Justin Yifu, Fang Cai, and Zhou Li. 2003. *The China Miracle: Development Strategy and Economic Reform*. Hong Kong: Chinese University Press.

Lipton, Michael. 1977. *Why Poor People Stay Poor*. London: Australian National University Press.

Liu, Shouying. 2017. "The Structure and Changes of China's Land Institutions." Paper presented at the 5th Young Scholars' Conference on China Studies, Baptist University of Hong Kong, Hong Kong, China, 19 December 2017.

Liu, Shouying. 2018. "Tudi Zhidu Biange Yu Jingji Jiegou Zhuanxing [Land System Changes and Economic Restructuring]." *Zhongguo Tudi Kexue [Chinese Land Science]* 32 (1): 1–10.

Liu, Yingjie. 1996. "Zhongguo Chengxiang Guanxi Yanbian De Lishi Fenxi [China's Rural-Urban Relations and Changes]." *Dangdai Zhongguo Lishi Yanjiu [Contemporary Chinese History Research]* 2: 1–10.

Logan, John. 2015. "China's Urban Future: The Difficult Road from Socialism." Paper presented at City and Society International Forum, 10-11 April 2015. Tongji University, Shanghai.

Lu, Yilong. 2004. *Chaoyue Hukou: Jiedu Zhongguo Huji Zhidu*. [Beyond Hukou: Interpreting China's Household Registration System.]. Beijing, China: Zhongguo Shehui Kexue Chubanshe.

Lui, Lake. 2016. "Hukou Intermarriages in China: Patterns and Trends." *Chinese Sociological Review* 49 (2): 110–137. doi:10.1080/21620555.2016.1231570.

Matthews, Mervyn. 1993. *The Passport Society-Controlling Movement in Russia and the USSR*. Boulder: Westview Press.

Meisner, Maurice. 1999. *Mao's China and After: A History of the People's Republic*. New York, NY: Free Press.

MGI (McKinsey Global Institute). 2006. "From 'Made in China' to 'Sold in China': The Rise of the Chinese Urban Consumer." November 2006. Accessed 10 January 2008. http://www.mckinsey.com/Insights/MGI/Research/Urbanization/From_made_in_China_to_sold_in_China

Ming, Holly. 2013. *The Education of Migrant Children and China's Future*. Oxon, UK: Taylor & Francis.

Moore, Mick. 1984. "Political Economy and the Rural-urban Divide." In *Development and the Rural-Urban Dividel*, edited by J. Harriss and M. Moore, 79–128. London: Frank Cass.

Naughton, Barry. 2007. *The Chinese Economy*. Cambridge, MA: MIT Press.

Normile, Dennis. 2017. "One in Three Chinese Children Faces an Education Apocalypse. An Ambitious Experiment Hopes to Save Them." *Science Magazine*, 21 September. Accessed

10 October 2017. http://www.sciencemag.org/news/2017/09/one-three-chinese-children-faces-education-apocalypse-ambitious-experiment-hopes-save

Oakes, Tim. 2019. "China's Urban Ideology: New Towns, Creation Cities, and Contested Landscapes of Memory." *Eurasian Geography and Economics* 60 (4).

Ong, Lynette. H. 2014. "State-Led Urbanization in China: Skyscrapers, Land Revenue and "Concentrated Villages"." *The China Quarterly* 217: 162–179. doi:10.1017/S0305741014000010. March 2014.

Palmer, James. 2015. "For American Pundits, China Isn't a Country. It's a Fantasyland." *The Washington Post*, 29 May. Accessed 20 October 2016. https://www.washingtonpost.com/opinions/for-american-pundits-china-isnt-a-country-its-a-fantasyland/2015/05/29/24ba60e0-0431-11e5-a428-c984eb077d4e_story.html?noredirect=on&utm_term=.d25795923c2f

Pan, Lu. 2018. "From Left-behind Children to Young Migrants: The Intergenerational Social Reproduction of Rural Migrant Labor in China." *Eurasian Geography and Economics* 59 (2): 184–203. doi:10.1080/15387216.2018.1484298.

Podrabinek, Alexander. 2013. "Revival of Serfdom in Russia." Institute of Modern Russia, 20 December. Accessed 23 December 2013. https://imrussia.org/en/analysis/politics/627-the-revival-of-serfdom-in-russia

Preobrazhensky, Eugenii. 1980. *The Crisis of Soviet Industrialization: Selected Essays*. White Plains, NY: M. E. Sharpe.

Pun, Ngai. 2005. *Made in China: Women Factory Workers in a Global Workplace*. London and Hong Kong: Duke University Press and Hong Kong University Press.

Pun, Ngai, and Jenny Chan. 2012. "Global Capital, the State, and Chinese Workers: The Foxconn Experience." *Modern China* 38: 383–410. doi:10.1177/0097700412447164.

Pun, Ngai, and Yan Ren. 2008. "Guojia Yu Nongmingong: Wufa Wancheng De Wuchanjiejihua [State and Rural Migrant Labor: Never Completed Proletarianization]." *21 Shiji Pinglun [Review of the 21st Century]* 6: 26–37.

Qi, Hao, and Zhongjin Li. 2019. "Giovanni Arrighi in Beijing: Rethinking the Transformation of the Labor Supply in Rural China during the Reform Era." *Science & Society* 83 (3): 327–354. doi:10.1521/siso.2019.83.3.327.

Qu, Xiaobo, Ge Yang, and Jie Cheng. 2015. "Renkou Liudong Qushi Yu Nongmingong Jiuye [Population Migration Trends and the Employment of Rural Migrant Labor]." In *Zhongguo Renkou Yu Laodong Wenti Baogao No.16 [Report on China's Population and Labor No. 16]*, edited by F. Cai and J. Zhang, 35–84. Beijing: Social Sciences Academic Press.

Ren, Yuan. 2017. "Liuyisi He Engesi Xianghu Duizhizhe De Dui Xiang Cheng Qianyi He Chengzhen Hua De Zhenglun [Debate on Lewis' and Engels' Views on Rural-urban Migration and Urbanization]." *Douban*, 15 October. Accessed 16 October 2017. https://www.douban.com/review/8866892/

Roach, Stephen. 2012. "Will Urbanization Save China's Capital Spending Bubble?" *Prudent Investor*. Accessed 22 October 2012. http://prudentinvestornewsletters.blogspot.hk/2012/08/will-urbanization-save-chinas-capital.html doi:10.1094/PDIS-11-11-0999-PDN.

Saunders, Doug. 2012. *Arrival City: How the Largest Migration in History Is Reshaping Our World*. New York: Pantheon Books.

SCDRC (State Council Development Research Center). 2014. *Cong Chengxiang Eryuan Dao Chengxiang Yiti Woguo Chengxiang Eryuan Tizhi De Tuchu Maodun Yu Weilai Zouxiang* [From Urban-Rural Dual Structure to Urban-Rural Integration]. Beijing: China Development Press.

Schmidt-Kallert, E., and P. Franke. 2013. *Livelihood Strategies of Multi-locational Households in The People's Republic of China*. Essen, Germany: Klartext Verlag.

Selden, Mark. 1993. *The Political Economy of Chinese Development*. Armonk: ME Sharpe.

Selden, Mark, and Jieh-min, Wu. 2011. "The Chinese State, Incomplete Proletarianization and Structures of Inequality in Two Epochs." *The Asia-Pacific Journal* 9 (5): No 1, 31 January 2011, http://japanfocus.org/-Mark-Selden/3480

Service, Robert. 2006. *Stalin: A Biography*. Cambridge, MA: Harvard University Press.

Shen, Mingmin. 1997. "Zhongguo Eryuan Shehui Jingji Jigou Yu Eryuan Shehui Tizhi Fenxi [The Dual Economic Structure and Dual Social System of China]." *Hong Kong Journal of Social Sciences* 9: 85–113.

Smith, R. E. F. 1968. *The Enserfment of the Russian Peasantry*. Cambridge: Cambridge University Press.

Solinger, Dorothy. 1999. *Contesting Citizenship in Urban China*. Berkeley, CA: University of California Press.

Stalin, Joseph. 1927. "Political Report of the Central Committee." Accessed 19 August 2017. https://www.marxists.org/reference/archive/stalin/works/1927/12/02.htm

State, Council. 2016. "'Yijian' Qiangdiao Shizhong Jianchi Nongcun Jiti Tudi Suoyouzhi De Genben Diwei ["The Opinion" Emphasizes Persistently Upholding The Fundamental Position Of Rural Collective Land Ownership]." 3 November. Accessed 17 October 2017. http://www.scio.gov.cn/xwfbh/xwbfbh/wqfbh/33978/35411/zy35415/Document/1515112/1515112.htm

Sun, Yan, and Chi Hu. 1997. "'Woguo Gongnong Chanpin Jiage Jiandaocha Wenti Fenxi [An Analysis of the Price Sisscisors Problem of Industrial and Agricultural Products in China]." *Difang Caizheng [Local Finance]* 5: 19–21 and 6 and 7: 13–14.

Tang, Anthony M. 1984. *An Analytical and Empirical Investigation of Agriculture in Mainland China, 1952–1980*, 4. Taipei, Taiwan: Chung Hua Institute for Economic Research, Mainland China Economic Series.

Tang, Wing-shing. 2014. "Governing by the State: A Study of the Literature on Governing Chinese Mega-cities." In *Branding Chinese Mega-Cities*, edited by P. O. Berg and E. Bjorner, 42–63. Chelternham, UK: Edward Elgar Publishing.

Tang, Wing-shing. 2019. "Town-Country Relations in China: Back to the Basics." *Eurasian Geography and Economics* 60 (4).

Tao, Ran, Fubing Su, Mingxing Liu, and Guangzhong Cao. 2010. "Land Leasing and Local Public Finance in China's Regional Development: Evidence from Prefecture-level Cities." *Urban Studies* 47: 2217–2236. doi:10.1177/0042098009357961.

Tiebout, Charles. 1956. "A Pure Theory of Local Expenditures." *Journal of Political Economy* 64 (5): 416–424. doi:10.1086/257839.

Todaro, Michael. 1969. "A Model of Labour Migration and Urban Unemployment in Less Developed Countries." *American Economic Review* 59 (1): 138–148.

Todaro, Michael. 1977. *Economics for a Developing World*. Hong Kong: Longman.

Treiman, Donald J. 2012. "The "Difference between Heaven and Earth": Urban–Rural Disparities in Well-being in China." *Research in Social Stratification and Mobility* 30 (1): 33–47. doi:10.1016/j.rssm.2011.10.001.

Tsui, Kai-yuan. 2011. "China's Infrastructure Investment Boom and Local Debt Crisis." *Eurasian Geography and Economics* 52 (5): 686–711. doi:10.2747/1539-7216.52.5.686.

Wang, Feiling. 2005. *Organizing through Division and Exclusion: China's Hukou System*. Stanford, CA: Stanford University Press.

Wang, Fenglong, and Yungang Liu. 2015. "How Unique Is 'China Model': A Review of Theoretical Perspectives on China's Urbanization in Anglophone Literature." *Chinese Geographical Science* 25 (1): 98–112. doi:10.1007/s11769-014-0713-2.

Wang, Helen. 2010. *The Chinese Dream: The Rise of the World's Largest Middle Class and What It Means to You*. Bestseller Press.

Wang, Haiguang. 2013. "Yizhi Yu Zhibian: Zhongguo Dangdai Huji Zhidu De Xingcheng Lujing Jiqi Sulian Yinsu De Yingxiang [Transplant and Involution: The Path Leading to the Present Hukou System in China and the Influence of the Russian Factor]." *Zhongguo Gaige Luntan Wang* [*Forum on Chinese Reform*], 15 May. Accessed 19 October 2013. http://www.chinareform.org.cn/society/huji/Experience/201305/t20130515_167130.htm doi:10.1094/PDIS-02-13-0213-PDN.

Wang, Jiye, and Jian Wang, eds. 1996. *Woguo Eryuan Jiegou Maodun Yu Gongyehua Zhanlue Xuanze* [The Conflicts of China's Dual System and the Strategic Options of Industrialization]. Beijing, China: Zhongguo Jihua Chubanshe.

Wang, Songji. 2016. *Zhongguo Chengxiang Shuangchong Eryuan Jiegou Yanjiu* [Study on the Double Dual Structure of Urban and Rural Areas in China]. Beijing, China: Renmin Chubanshe.

Wang, Yu, and Christine, R. Schwartz. 2015. "Intermarriage and Social Exclusion in China." Center for Demography and Ecology University of Wisconsin-Madison, Working Paper No. 2015-01.

Wantanabe, Susumu. 1994. "The Lewisian Turning Point and International Migration: The Case of Japan." *Asian and Pacific Migration Journal* 3 (1): 119–147.

Webber, Michael. 2008. "The Place of Primitive Accumulation in Rural China." *Economic Geography* 84 (4): 395–421. doi:10.1111/j.1944-8287.2008.00002.x.

Webster, Chris. 2014. "Conclusion." In *Rural Migrants in Urban China: Enclaves and Transient Urbanism*, edited by F. Wu, F. Zhang, and C. Webster, 277–292. Oxon, U.K.: Routledge.

Wei, Yanning, and Yue Gong. 2019. "Understanding Chinese Rural-to-Urban Migrant Children's Education Predicament: A Dual System Perspective." *International Journal of Educational Development*. Online first. doi:10.1016/j.ijedudev.2019.05.001.

Wemheuer, Felix. 2014. *Famine Politics in Maoist China and the Soviet Union*. New Haven and London: Yale University Press.

Whiting, Susan. 2011. "Values in Land: Fiscal Pressures, Land Disputes and Justice Claims in Rural and Peri-urban China." *Urban Studies* 48: 569–587. doi:10.1177/0042098010390242.

Whyte, Martin K. 2010a. "Social Change and the Urban-Rural Divide in China." In *The Irish Asia Strategy and Its China Relations*, edited by F. Hong and J.-C. Gottwald, 45–60. Amsterdam: Rozenberg Publishers.

Whyte, Martin K., ed. 2010b. *One Country, Two Societies*. Cambridge, Massachusetts: Harvard University Press.

Wissink, Bart, Arjan Hazelzet, and Werner Breitung. 2014. "Migrant Integration in China: Evidence from Guangzhou." In *Rural Migrants in Urban China: Enclaves and Transient Urbanism*, edited by W. Fulong, F. Zhang, and C. Webster, 99–120. Oxon, U.K.: Routledge.

Wong, Bin. 2008. "Transformations of China's Post-1949 Political Economy in a Historical Perspective." *Pacific Economic Review* 12 (3): 291–307. doi:10.1111/j.1468-0106.2008.00402.x.

Wu, Fulong. 2008. "China's Great Transformation: Neoliberalism as Establishing a Market Society." *Geoforum* 39 (3): 1003–1096. doi:10.1016/j.geoforum.2008.01.007.

Wu, Jieh-min. 2017. "Migrant Citizenship Regimes in Globalized China: A Historical-Institutional Comparison." *Rural China: an International Journal of History and Social Science* 14: 128–154.

Wu, Xiaogang, and Bingdao Zheng. 2016. *Household Registration, Urban Status Attainment, and Social Stratification in Contemporary Urban China*. University of Michigan Population Studies Center Research Report 16–874.

Xinminwang. 2014. "Huanhu Chengxiang Eryuan Huji Zhidu Shouzhong Zhengqin [Hurray to the Demise of the Rural-Urban Dual System]." 18 August. Accessed 20 August 2014. http://finance.ifeng.com/a/20140818/12948010_0.shtml

Xiong, Yihan. 2015. "The Broken Ladder: Why Education Provides No Upward Mobility for Migrant Children in China." *The China Quarterly* 221: 161–18. doi:10.1017/S0305741015000016.

Yang, Dennis, and Fang Cai. 2003. "The Political Economy of China's Rural–Urban Divide." In *How Far across the River? Chinese Policy Reform at the Millennium*, edited by N. Hope, D. Yang, and M. Yang, 389–416. Stanford. CA: Stanford University Press.

Yang, Dongping. 2016. "Zhongguo Liudong Ertong Jiaoyu De Fazhan He Zhengce Yanbian [Evolution and Changes of China's Migration Children Education Policy]." In *Zhongguo Liudong Ertong Jiaoyu Fazhan Baogao* [Changes of China's Migration Children Education Policy], edited by D. Yang, 1–20. Beijing: Shehui Kexue Wenxian Chuban She.

Yang, Junshi. 1987. *Xiandaihua Yu Zhongguo Gongchan Zhuyi* [Modernization and the Chinese Communism]. Hong Kong: Chinese University Press.

Yang, Jisheng. 2012. *Tombstone: The Great Chinese Famine, 1958-1962*. (translated by Jian Guo, and Stacey Moshe). New York: Farrar: Straus and Giroux.

Yang, Juhua. 2015. "Renkou Liudong Yu Juzhu Fenli: Jingji Lixing Yihuo Zhidu Zhiyue? [Population Movement and Residency Separation: Rational Economy or Institutional Constraint?]." *Renkou Xuekan [Population Journal]* 37 (1): 362–391.

Yao, Rosealea. 2014. "The Slums are Still with Us." *Ideas*. Gavelal Dragonomics. 4 September. doi:10.1128/AEM.03322-14.

Ye, Jingzhong. 2018. "'Sannong Wenti': Bei Kuada De Xueshu Gainian Jíqi Juxian." ['Three Rural Issues': An Exaggerated Academic Concept and its Limitations.] Dongnan Xueshu *[Southeast Journal]*. 5. 1 November. Accessed January 30. https://mp.weixin.qq.com/s/_337z9b-U3ZtmmNXd0H_9A

Ye, Jingzhong, Chunyu Wang, Huifang Wu, Congzhi He, and Juan Liu. 2013. "Internal Migration and Left-behind Populations in China." *The Journal of Peasant Studies* 40 (6): 1119–1146. doi:10.1080/03066150.2013.861421.

Ye, Jingzhong, Huifang Wu, and Xiangdan Meng. 2015. *Zhongguo Nongcun Liushou Renkou Fansi Fazhan Zhuyi De Shijiao [Left-behind Population in Rural China:A Critique of Developmentalism]*. Beijing: Social Sciences Academic Press.

Yu, Jianrong. 2002. "Nongmin Ruhe Shiqu Tudi Suoyouquan [How Farmers Lost the Ownership of Land]." Accessed 29 August 2017. http://www.usc.cuhk.edu.hk/PaperCollection/Details.aspx?id=1754

Yu, Zhongying. 1994. "Lun Zhonguo De Erji Jingji Ji Yingxiang [China's Bi-polar Economy and Its Impact]." *Fudian Xuebao [Fudan Journal] (Social Sciences Edition)* 2: 21–26.

Zaslavsky, Victor, and Yuri Luryi. 1979. "The Passport System in the USSR and Changes in Soviet Society." *Soviet Union* 6 (Pt .2): 137–153.

Zhan, Shaohua. 2017. "Hukou Reform and Land Politics in China: Rise of a Tripartite Alliance." *The China Journal* 78: 25–49. doi:10.1086/690622.

Zhang, Li, Richard LeGates, and Min Zhao. 2016. *Understanding China's Urbanization: The Great Demographic, Spatial, Economic, and Social Transformation*. Cheltenham, UK: Edward Elgar.

Zhou, Qiren. 2017. *Chengxiang Zhongguo* [Urban-Rural China]. Beijing: Zhongxin Publishing Group.

Zhou, Yu, George CS Lin, and Jun Zhang. 2019. "Urban China Through The Lens of Neoliberalism: Is a Conceptual Twist Enough?." *Urban Studies* 56 (1): 33–43.

Zhu, Yu, Liyue Li., et al. 2014. "From 'integration into Cities' to 'an Integrated Society': Women Migrants' Needs and Rights in Fujian Province, China." In *Migration, Gender and Social Justice: Perspectives on Human Insecurity*, edited by T. D. Truong, et al., 153–171. Hexagon Series on Human and Environmental Security and Peace 9. Heidelberg, Germany: Springer. https://link.springer.com/content/pdf/10.1007%2F978-3-642-28012-2.pdf

Town-country relations in China: back to basics

Wing-Shing Tang

ABSTRACT

This paper develops a nuanced understanding of town-country relations in China. Inherent in the current literature is the serious problem of randomly indigenizing experiences and concepts from the West. To restore this imbalance, this paper provides a preliminary attempt to develop a better understanding methodologically, epistemologically and theoretically. It argues first for the implementation of a spatial story methodology that takes spatio-temporal relations seriously; second, a nuanced concept of the state as dynastic empire-building which functions through a two-tier governing relationship in focus-field relationships; third, a rejection of dualism, advocating instead the idea of mutual embeddedness in order to decipher the subtle relationships between parts and whole; fourth, an interrogation of the appropriateness of spatial concepts like spatial reach, scale, level and territory; and, finally, a proposed concept of *cheng*-cum-*xiang* to eschew a dualistic approach to town-country relations. In conclusion, the paper suggests implications for debates within urban studies more generally.

Introduction

In an earlier paper (Tang 2014a), I criticized the literature on urban China for the proliferation of random conceptual indigenization and random conceptual appropriation from the West. Due to these numerous random attempts, the literature has not interrogated urban China properly. The crux of the matter lies in the philosophy and methodology underlying burgeoning conceptualizations of urban China. In their random indigenization and appropriation, many urban China researchers do not recognize the western dualism that informs these western concepts. Worse still, during indigenization and appropriation, they have often discarded the dialectical dimension by exaggerating the prevalence of two somewhat disparate forces, paths, trends or movements, only in relation with each other in unwitting occasions. Accordingly, there are in the literature all sorts of dualisms, including center-local, plan-market, state-society, town-

---

This article has been republished with minor changes. These changes do not impact the academic content of the article.

country and state-collective land ownership, yet with meager discussion of the interaction between these polar opposites. For example, there is, on the one hand, state-led industrialization, urbanization, gentrification, place-making, etc. Although the dominance of the state in producing these urban processes and patterns has been repeatedly acknowledged, discussions, which have been simplified by deploying the signifier "state-led", usually lack the fine details of how the state really dominates and the delicate inputs of, and interaction with, other non-state agents. On the other hand, in the discussion of similar patterns in the countryside, the ideas of the dual-sector economy, dual-track urbanization, etc. have exaggerated the operation of village-led forces, usually portrayed as being independent of the state. These formulations would have us believe that state-led urban phenomena could be understood independent of the village-led ones or, more seriously, *vice versa* (Wong 2019). The relationship between the state and urban space is too complex to be captured by these non-dialectical, dualistic conceptualizations.

It is the objective of this paper to provide some analytical strategies aimed at correcting the problem in the literature, with special attention to town-country relations. The second section summarizes how these relations have been depicted in the literature. Until recently, the tradition has considered town and country as separate. One may attribute this to the sociologist's invention of the rural tradition in the understanding of Chinese society more than a century ago. The historians complement it by arguing that being a rural society, the rural-urban distinction in the West was something external to China. Later, once settlements like cities in the West started to emerge, "the Chinese city" was conceptualized as modern, and the rural as residual and traditional. This was the rhetoric of the urban-rural dichotomy. Since the establishment of the People's Republic of China and the implementation of the *hukou* registration and rationing systems, people tend to emphasize the dichotomy between town and country. Due to the recent developments in both industry and agriculture, and in regional administrative reforms in both the urban and rural systems, academics have wasted no time in following the state by updating their conceptual apparatuses and labeled the phenomenon "urban-rural integration," as if the conceptual dichotomy between town and country has disappeared. The third section reviews some of the attempts that have taken the Chinese situation more seriously. This work tends to suffer from ahistorical generalization, extrapolating the present system of *shi* backward into the dynastic past, or historicized simplification, downplaying the Chinese politico-economic dynamics.

This survey of the literature calls for a complete revamping of our understanding of town-country relations from integrated epistemological, methodological and theoretical perspectives. The fourth section puts forth the methodology of spatial story. Informed by *tongbian* philosophy, it argues for comprehending town-country relations in the process of becoming. The relationship between the emperor and his society consisted of many mutually embedded forces, which can be represented as focus-field relationships. Due to limited space, the theorization of town-country relations can only be limited to the dynastic past. The fifth section coins the concept of dynastic empire-building (DEB). Unlike the modern state, the Chinese empire was interested in fostering an agrarian commercial economy and establishing one jurisdiction: that of the throne. The spatialization of DEB resulted in the establishment of a two-tier governing system – *xian* on the one hand, and, on the other, its intermediate superiors. These two tiers can be understood as constituting complicated, yet dynamic, focus-field relationships. Accordingly, *xian* was a space of dependent delegate and *cheng* was the administrative seat within it. *Cheng* and *xian* are understood as two polarities always in the process of becoming and continuing. Underneath this process was the penetration and orchestration by the empire. It then becomes obvious that town-country relations are better understood as *cheng*-cum-*xiang* relations. The sixth section continues the discussion by focusing on the evolution of Guangzhou *cheng*. The final section attempts to draw on some of the implications of this formulation for urban China research as well as urban research in the world.

Randomized town and country dualities

> [C]ity/country has never been a significant Chinese distinction.
> (DeGlopper 1995, 43)

> In the administrative system,
> our country since the Qin and Han dynasties has already employed
> the system of '*cheng*-cum-*xiang* administration (*chengxiang hezhi* 城鄉合治),' …
> the '*shi* 市' designation has almost covered every single inch of land in the whole country.
> (Sun 2005, 10, my translation)

Given Marco Polo's account in 1276 of magnificent city life in what we today call Hangzhou, and the dominance of agricultural activities in China for more than 3,000 years, it is surprising to remark that the issue of town-country relations has only been more intensively studied in the past forty years or so. On the one hand, reforms both in agriculture and in the rural areas have called for the elimination of the rural-urban gap. On the other, reforms in urban industry and in regional administration have emphasized the promotion of urban-rural integration (*chengxiang yitihua* 城鄉一體化) (e.g., Kipnis 2013; Tang 2015; Zeuthen and Griffiths 2011; Zhou 1997). The consensus of this literature is that while the rural-urban dichotomy prevailed during the Maoist period, there has been a softening of this divide during the reform period. To interrogate this long negligence and recent mis-focus, it is necessary to explain how town-country relations have been conceptually defined, what sort of concepts have been invoked to unravel the underlying dynamics, and the extent to which these concepts are perceptive at all.

The rural tradition invented

One can find a clue to the near absence of earlier studies on town-country relations in Liu's (2002) review of the development of urban anthropology in China. The literati tradition of Chinese scholarship in the dynastic past did not favor the study of real lives of real people. Since the turn of the 20th century, thanks to the efforts of various generations of Sinologists as well as Chinese scholars, the assumption that Chinese society was fundamentally rural in nature had started to take root. Social life was studied through the observation of rural families in the village; then, the village was considered the real setting to study society; and, finally, the rural nature of society was to be theorized by lineage. Many of these ideas were basically drawn from the West, at the time when people encountered modernity. This applied to Sinologists such as Maurice Freedman, who relied on his British Africanist colleagues working in the African context for his lineage paradigm, or Fei Xiaotong, who drew on Robert Park and the Chicago School for the areal unit to understanding real life in China. The dualism between modernity and tradition, originally invented in the West, has since been indigenized in China. That China was considered backward gave the rural a conceptual supremacy in the representation of Chinese society. Not surprisingly, the urban setting was assumed to be rural in nature too, as DeGlopper's (1995, 45) study of Lukang, a city in Taiwan, has illustrated: "urban is an adjective rather than an essence," and "[c]ities are just as 'natural' as villages."

A corollary to this thesis was that the town-country dichotomy was considered absent in imperial China. As usual, the China urban history literature (e.g., DeGlopper 1995; Faure and Liu 2002; Lewis 2006: 135–188; Mote 1977; Xu 2000: 77–94, 158–165) starts with an interrogation of existing findings about cities in the West, including Louis Wirth's urbanism as a way of life in early 20th century Chicago, Robert Redfield's rural-urban continuum, Raymond Williams' continual changing qualities of the idea of the country from that of the city, and Lewis Mumford's rise to supremacy of a single great metropolis upon the creation of despotic states in Europe. In comparison, Chinese civilization has not relied on "a self-identifying and self-perpetuating urban elite as a component of the population" (Mote 1977, 102). This backdrop allows Mote (1977, 105) to affirm that "[t]he rural . . . defined the Chinese way of life". Based on studies on Nanjing and Suzhou (Mote 1977: 119; 1973), he has found that the urban and the rural did not differ much on dimensions such as "the physical form of cities," "styles in architecture and dress," "urban and rural attitudes in elite (and perhaps in popular) psychology," "the structure and character of cultural activities," and "some glimpses into the pattern of economic life." In Mote's conclusion, there was an urban-rural continuum in imperial China. Faure and Liu (2002, 2–3) once challenged Mote by quoting the experience of (commercial) towns like Hankou in the past, where traders gathered in merchant enclaves, forming commercial centers not too dissimilar to many western cities. Nevertheless, at the end, they come to a somewhat similar conclusion that the "rural-urban distinction was external to traditional Chinese thought and is even now, only partially integrated into Chinese society," (Faure and Liu 2002, 14).

The city as agent of modernization

Since the Republican era, urban scholars and practitioners in China were bothered by the relationship between modernity and national identity. To struggle for a place in the world, they looked to the city, as Esherick (2000) has noted:

> In China in this period, with a vast countryside and a weak central government, the city also proved to be a manageable social and political unit for significant modernizing efforts.

Study after study has enumerated for us the usual material artifacts of being modern in these cities – concrete roads, modern trams, railways, water pipe supply, cables, electricity, telephones, public parks, department stores, printing, etc – and the possible application of scientific knowledge to form the modern art of governing. This literature has emphasized a stable and proper order, without, however, paying heed to, in the Foucauldian term, the penetration to the level of everyday life as well as the conversion of *min* (民 subject of the sovereign) to *qunzhong* (群眾, "masses") or *shehui* (社會, "society"), the domain of modern government (e.g., Tsin 1999).

This scholarly tradition of treating the city as modern and the countryside as traditional has basically been followed even after the Chinese Communist Party took the reins. The most representative in the literature is the economic approach, which may be traced all the way to modernization theory reiterating the modern-tradition dualism at a time when a considerable number of colonies started to gain political independence after WWII. Development economists differentiate two sectors in the developing economies (the Lewis model): the modern industry and the backward agriculture. More specifically, extensive industrialization will attract surplus labor from the agricultural sector to the modern industrial sector in the city until it is exhausted. Chan (1994) expands on this with a socialist twist by incorporating the Soviet experience of economizing urbanization costs. In other words, it is the economic strategy of the day that has accounted for the imposition of *hukou* (戶口) registration system and associated measures, the "invisible walls," to restrict surplus labor in the countryside from migrating to the city and, in corollary, the consolidation of the rural-urban dual system. In this special issue, Chan and Wei (2019) challenge the relevance of the Lewis model to China, pointing out that his argument is premised on a single society while the Chinese dual system is rested on two separate societies. This system renders migrant labor cheap for the world's factory and allows the state to expropriate rural land.

This rather economistic argument has, however, been challenged by the historico-cultural approach, as exemplified by Lin (2006, 79, 61). Lin emphasizes that the "virtual urban-rural segregation" was caused not so much by the Soviet-inclined economic development strategy *per se* as by a state-centered intermediation of nationalism, socialism and developmentalism. She argues that China's path to modernity belongs to "history 3," that is, neither the western world's history 1 nor the colonies' history 2. Although China has also emphasized development, it is the strong revolutionary flavor that differs it from colonial modernity. Although it adheres to socialism, it departs from Stalinist statism. While it started to integrate with the capitalist world thirty or so years ago, it seeks many reforms that are absent in those countries. Unlike developmental states in the developing world, the powerful socialist state is capable of penetrating society and monopolizing power because of "a societal consensus, a social contract, furnished by the revolution and the party's long-term mobilization at the grassroots," (Lin 2006, 70). The observable urban bias must be explained by China's external relations with global and regional Cold War contenders during early socialism, instead of a simple equation with the Soviet strategy, while rural collectivization was implemented for different reasons and with diverging effects. China did not practice the "worker-peasant alliance" approach, thereby allowing the rural areas to play a major part in determining their own economic development. In addition, it did not recruit village cadres from the cities, thereby continuing to foster traditional rural values (Tang 1997, 41). Furthermore, the state had great interest in changing its relationship with the agrarian population; Liu Shaoqi once remarked that "the future of socialism relies on not only a proletarian state and a big industry but also cooperatives" (Lin 2006, 79). These enlightening, complex processes aside, however, the historico-cultural approach does not deviate much from the economic approach in the dualistic understanding of the town as modern and the countryside as traditional.

The randomly indigenized Chinese city

Urban processes have also been frequently misinterpreted. One example would be Lin's (2007) analysis of urbanism in China, which has indigenized several fashionable concepts (see also McGee et al. 2007, 29–52). Lin sees it as necessary to conceptualize urbanism in China within the framework of state-society relations. More specifically, urbanism, which is defined as the ideological and material processes of social change (the latter in forms of spatial mobility), is an integral part of the state scaling strategy to pursue national development agendas. This thesis is elaborated historically, although the bias is clearly tilted toward the socialist regime. Under Mao, a distinct mode of urbanism, usually portrayed as anti-urbanism, was adopted. It was a strategy of urban containment, supported by the household registration system introduced since 1958, to expedite industrialization and urban manageability. The "invisible walls" drew up a sharp distinction, both ideologically and materially, between town and country and erected a blockage of direct, physical interaction between the two. During the reform period, the neoliberal state practiced a more relaxed and flexible state-society relation according to market forces, allowing for a significant degree of freedom in the movement of people, resources and activities. As a result, urbanism has grown by leaps and bounds, linking town with country. In his discussion, there is an acknowledgment by Lin (2007) of the relevance of Lefebvre's theory: "[i]n this regard, Henri Lefebvre's (1991) 'conceptual triad' of 'the representations of space ..., 'space of representation' and 'spatial practices' provides an interesting framework for analyzing."

This kind of random conceptual indigenization, obviously, invites criticism. Lin has adopted the state-society concept coined elsewhere, without scrutinizing the debates about the applicability of civil society and nation-state to China (e.g. Duara 1997; Huang 1993; Ma 1994; Solinger 1993). The indigenized concept of neoliberal state is also inappropriate for deciphering the Chinese government in the recent decades. Since Tang (2014a, 51–53) and Buckingham (2017) have already argued emphatically against it elsewhere,[1] it is unnecessary to repeat it here. Moreover, the Lefebvrian concepts are usually mentioned in passing, if not merely by name. We are totally kept in the dark as to why, how and in which context Lefebvre is relevant for analyzing not only state-society relationship but also the growth and spatiality of urbanism in China. Has abstraction occupied a prominent position in the Chinese intellectual history? Has space in China taken up a role similar to its western counterpart dated back even to the *polis* (see below)?

Random conceptual indigenization appears even in the basic concept of the city and its cognates. Faure and Liu (2002, 2) make a penetrative comment on the concept of the city:

> In Chinese, the distinction of the cities such as Nanjing and Suzhou from the countryside was, and still is, often expressed in terms of the opposites *cheng* (city) and *xiang* (village or group of villages). Writing about China and Chinese concepts in English carries its own pitfalls, for the conventions of translation do not adequately describe cross-cultural differences; what the educated Westerner might mean by 'city' or 'village' does not necessarily correspond to what the educated Chinese person means by *cheng* or *xiang*. At the heart of Mote's papers, therefore, is the important reminder that China has to be understood on its own terms.

Sun (2005, 10, my translation), whose quotation was cited earlier, has aired a similar concern: "the western concept of city and the characteristics of city as well as its functions as depicted in various theoretical discussion do not exhibit in our '*chengshi*'".

One finds similar mistakes regarding the concept of suburb, which has been widely employed out of context. Even in imperial China, we are told by Lin (2007, 14, 16, 20, my emphases) that the suburbs "were favored sites for markets and businesses serving the rural populace," and that, "By the nineteenth century, few walled cities lacked *suburban* development outside at least one gate, and in many cities, particularly those in Guangdong, built-up areas in the *suburbs* exceeded those within the walls." During the Maoist regime, there "was little urban-rural interaction in either the *suburban* or peri-urban regions." In the succeeding neoliberal period, "millions of urbanites are moving out of the urban centers to the *suburb*." Lin's widespread random indigenization of the concept of suburb must have overlooked the warning by historian Xu (2000, 164, emphasis original): "the term *suburb* used here may be somewhat misleading if its Western concept were to be rigidly applied to the Chinese state of affairs." For sure, the suburb in the West was developed from industrial capitalism in the city. Due to the proliferation of urban problems within the inner city, the middle and upper classes fled to the suburb, leaving behind the working class. Accordingly, a modern suburb "is essentially in a dependent relationship to the whole organism of the city, and the complete suburban area of a city performs only part of its total functions," (Dyos, quoted in Xu 2000, 164]). The Chinese countryside, *jiaoqu* (郊區), was not a suburb in the western sense. There was no dependent relationship to the city in the past, as reflected in Xu's study of *Xijiaoqu* of Suzhou in imperial China, which was more central in terms of regional trade and business and bustling daily life, and more "urban" than most areas within the city walls. Villages have been included in *chengshi* as *jiaoqu* since 1949 to guarantee the production of food staples for the people and raw materials and resources for industrial production, but with different land relations. It is on the basis of Faure and Liu's earlier remark "that China has to be understood on its own terms" that we now turn to those analyses that have built on more indigenous understanding of China.

Simplified town and country dualities

Ahistorical generalization

The anchorage point for this intervention is the confusion caused by Sun's earlier quotation that the *shi* designation has covered almost every single inch of land in the whole country. This is not the case on two grounds. *Shi* as a designation was a comparatively recent invention. Preliminary forms of an independent spatial boundary together with independent administrative offices may be traced back to the Song dynasty, but the version comparable to the West had only started in the Republican era, while the current version was implemented after the Chinese Communist Party took the reins (Han, Lin and Wei 2011; Liu and Fan 2015, 99–150; Liu, Ma, and Xiong 2012, 26). In contrast, the administrative system of *cheng* has been employed since the dynastic era. Besides, Chinese history informs us that *cheng* had never covered "every single inch of land." Instead, it was *xian* (縣 county) that has been the most basic local unit in the spatial administrative hierarchy since 514 B.C. (Zhou 2005, 28). This type of ahistorical statement formulated from thin air is very popular nowadays when Party propaganda is in full swing. While it is imperative to build our understanding on the reflection of more indigenous details, it is absolutely unacceptable to do so on unsound, backward extrapolation.

Historicized simplification

In their theorization of the urban administrative district, Liu and his associates (Liu, Ma, and Xiong 2012, 31) have put forward the distinction between *diyuxing zhengqu* (地域型政區 regionally delineated administrative districts) and *chengshixing zhengqu* (城市型政區 city delineated administrative districts). The former represents a widespread, areal distribution of mainly agricultural productive forces with a small urban built-up area, whereas the latter refers to a concentration of secondary and tertiary industries, forming a large built-up area. They further generalize the two as representing the morphology of *chengxiang hezhi* (城鄉合治 town-country integrated governance) and of *chengxiang fenzhi* (城鄉分治 town-country discrete governance), respectively. In historical terms, the dynastic past and the post-reform situations exhibit the former characteristics, whereas Maoist China represents the latter (see also Liu and Fan 2015, 22–31). This conceptual distinction is justified on the basis of the level of productive forces: once the growth of cities and towns, due to the inducement of economic development, had attained a certain threshold, it became necessary to establish independent administrative units to handle these burgeoning urban affairs. This conjecture has been justified on the basis of world historical development. The real issue underlying the development of cities in the West, however, is not so much one of quantitative change leading to qualitative change as one of the presence or absence of autonomy. This is also the rationale underlying the administrative formation of many metropolitan areas all over the world. The randomly indigenized, and rather crude, Marxist development argument – the imperative of the development of the productive force – by Liu and his associates has basically downplayed the Chinese politico-economic dynamics.

Summary

The above review has demonstrated that understandings about town-country relations in China have heavily indigenized concepts and experiences from the West. As a matter of fact, why do they in general make no reference to a closer partner like the Soviet Union, from which China had learnt a lot during its early days of socialism? What were the concrete experiences in those days? It is common for the literature to treat these experiences simplistically as if they were discrete from the post-reform development, as well as not essential to it at all (Tang 2014a, 43–44).[2] Besides, of utmost importance is the legacy of the dynastic past. China historians nowadays have repeatedly urged us to understand the present on the basis of the past. This advice certainly applies to the current discussion on town-country relations. All these diachronic and synchronic interactions must be elaborated in terms of geography of power and politics too.

Eschewing randomness: the methodology of spatial story

Methodologically, we need a framework that allows us to interrogate town-country relations within the context of the interaction between China and the rest of the world as well as that between China's past and present.

Elsewhere (Tang Forthcoming), I draw on *tongbian* (通變) philosophy (Tian 2005) – itself developed from *yin-yang* philosophy – to characterize the paired relationships between China – "here" – and the world – "there" – and between the past and the present as mutually embedded. Each pole of the pair grows from the other and needs the other as partner. They are in ceaseless interaction, implying the process of becoming, continuity and change. Since everything is in the process of change, it is better to conceive of it as a "focus-and-field" relationship. While the focus represents a place of convergence, the latter refers to the sphere of influence from the perspective of the focal center. As both foci and fields are never finally fixed, their boundaries are always changing over time. It is then difficult to equate the field with the familiar one-many and part-whole models; in contrast, the focus-field perspective employs a "this-that" model. Given the great variety of specific contexts that constitute the field, it is imperative to detail the context in order to specify the focus. Put differently, by indulging in local and temporal investigation, we would be able to find out the viable contexts that constitute the focus and that are in turn constituted by them (Hall and Ames 1995, 272–78).

To operationalize it, I propose that it is the mutual embeddedness between space and time that we need to unearth, and that this can be achieved by the methodology of writing a spatial story. The latter involves narrating a subject's process of becoming. It regresses from the present to the past so as to progress from the present to the future. In recognizing the importance of becoming diachronically, the methodology underscores its synchronic dimension as well: its interdependence among its various internal components as well as with external others at any one period. The former includes the relations between Hong Kong and the mainland, while the latter refers to Britain's relation with Hong Kong as well as the Soviet Union with Chinese socialist development. For the former, China was affected by the various ways that Britain had gone through mercantilism (on the body), early (colonial production of space) and late capitalism (urban production of space). For the latter, Soviet Union's development as a vanguard state is influential. All these interdependences are spatially mediated in that the resources invoked in consolidating or perpetuating the intermediation vary in spatial reach. It is the spatio-temporality of the subject, always in the process of becoming, that a spatial story attempts to unveil (see Figure 1).

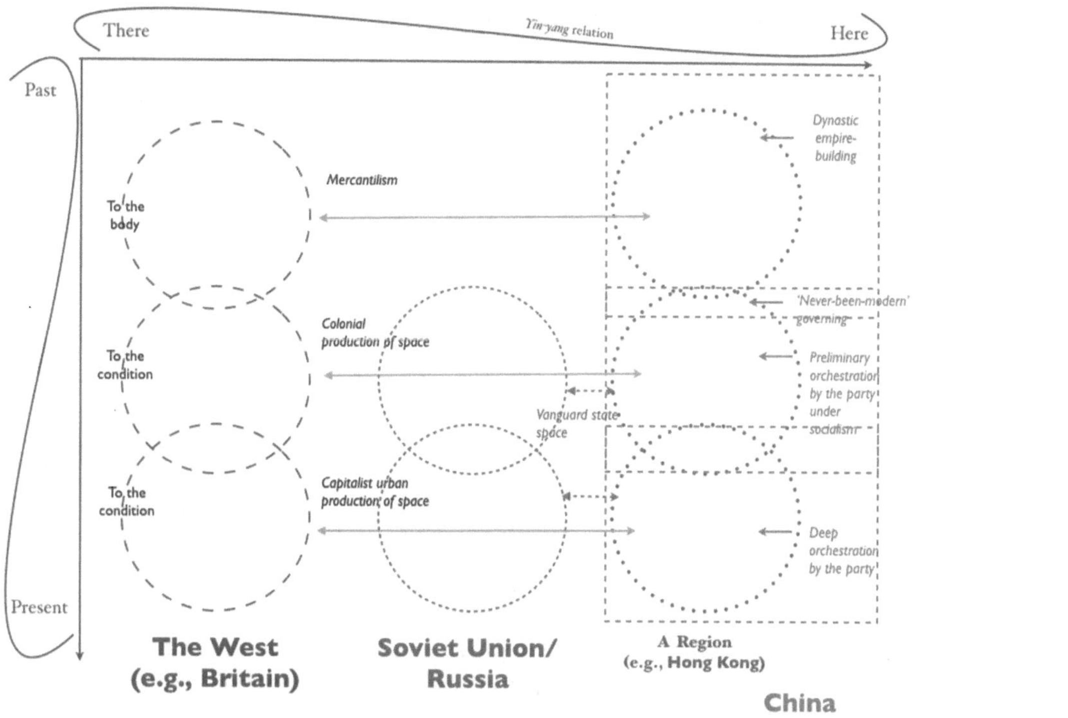

Figure 1. The spatial story approach (Source: modified from Tang (Forthcoming).

It is this methodology that informs our approach to town-country relations in China. Accordingly, the town-country relation is the subject in the process of becoming. What it is today has its historical trajectory. It has developed from its relationship with various other components in the Chinese society in the past. These relationships over time help inform and mold the succeeding possible nature and contents of town-country relation. Besides, as China is in constant interaction with the outside world, its town-country relation also incorporates into it the interrelationships between China and its counterparts elsewhere. These incorporations can take the form of knowledge production as well as material intervention. It is these spatio-temporal interdependencies over time, continuously changing and in the process of becoming, that have defined town-country relations. In particular, the relationship between the emperor and his society consisted of many correlates, which can be understood as some kind of focus-field relationship. Within the field, there are towns and countries, among others, which are themselves foci and fields. Thus, town-country relations can also be interpreted by focus-field relationships.

Informed by this methodology, the rest of the paper will unravel the making of town-country relations in China. Due to the limited scope of this paper, we have to focus only on relations during the dynastic past, while leaving those during the Republican era and Maoist and post-reform China to later research. As focus-field relationships vary within any focus, it is imperative for us to further elaborate town-country relation during the dynastic past by immersing ourselves in a particular city: Guangzhou.

Cheng-cum-*xiang* under dynastic empire-building

The *raison d'être* of DEB

> When we take European developments as the norm,
> all other experiences appear to be abnormal. ...
> For economic dynamics specifically, ...
> we tend not to be aware of, let alone search for,
> dynamics of change that do not fit
> within a narrative of capitalist development.
> (Wong 1999, 210, 211)

Dynastic empires, which differed from nation-states in the West, had their own concerns. Unlike commercial capitalism, the agrarian commercial economy of the empire relied on tilling new land while repairing and expanding water control and intensifying usage on existing land to produce grain. As the economic welfare of the people was conceived as basic to the stability of the empire, its two particular interests were grain commerce and rural industries. There was a concern for long-distance grain trade, discouraging the hoarding of grain and the concomitant agglomeration of merchant power. In addition, rural industries were perceived as a supplemental income for the household economy so as to bolster the economic viability and social cohesion of the family. As a result, two types of agrarian economy emerged: a series of small-scale, self-sufficient economies everywhere, together with a complex large-scale interdependent economy monitored by the imperial administration. Thus, unlike commercial capitalism, there was no market centralization forcing the integration of the countryside and large cities as well as no dominance of rich merchants in fiscal security and political power and legitimacy (Wong 1999).

Ever since the Zhou dynasty, every emperor, whose rule was "all-under-Heaven", attributed his legitimacy to Heaven. He was the only one responsible to Heaven, while everyone and everything else was solely dependent upon his will and largesse. To affirm the heavenly mandate, he was interested in establishing one jurisdiction – that of his throne (Huff 2003, 259). While aligning himself with the forces of nature, he was interested in subjecting all his subordinating subjects to follow the prescribed path and submission to the will of the throne. There was then a hierarchy between the emperor and the rest, with the former being always the creator and the latter the created.

Spatialization: a two-tier DEB system in focus-field relationship

To facilitate empire-building, the emperor must institutionalize a spatialization schema to perpetuate the hegemony of one jurisdiction while promoting the agrarian economy. First, unlike the sovereign states in Europe, as argued by Isin 2007, 216), it was difficult, as well as inappropriate, to represent space as an abstraction in Lefebvrian terms. The tradition of the *logos* in the West is interested in searching for causal laws and formal patterns. Underneath the concern for truth is, dating back to Plato, the dualism between truth and non-truth. Later developments, more specifically since the Enlightenment, include Descartes' distinction between mind and matter. One view of space – Cartesian space – had since then started to dominate western thought. It is an abstract, mental construction, to be conceptualized rationally. In contrast, classical Chinese culture practiced *ars contextualis* rather than *ontologia generalis* and *scientia universalis*. It presumes no essential features, favoring instead the elaboration of how things are harmoniously correlated. The absence of the dualism between reality and appearance calls for the articulation of one and the other by the symmetrically related concept of polarity. The practice of *li* (禮) – the sacred rites of the past – had further prevented rational reasoning and held abstraction in disfavor. Way-seekers – as Hall and Ames (1987, 105) call these practitioners – mostly "search out those forms of action that promote harmonious social existence". Unlike their counterparts – truth-seekers – they have no interest to conceptually characterize the way things are. Accordingly, instead of being an abstract conception, space during the empire was a down-to-earth, concrete context that situates the correlation of events.

Second, the emperor would not practice a technique of spatialization assuming individuality and its associated quest for autonomy (as publicized by Foucauldian scholars of governmentality). According to Aristotle, *polis* was a separate spatial container where people did politics while eschewing daily chores. Then, in 12th and 13th century Europe, legal autonomous collectives, including cities and towns, emerged as neutral space, relatively free from the interference of religious and political censors (Huff 2003, 251). A few centuries later, states tried to establish their sovereignties by eliminating legal autonomous bodies other than individuals. A series of liberal reforms in the early 19th century dramatically altered the concept of municipal government with the clear concern about governing cities. This brief historical account illustrates that thanks to the various quests for autonomy, the city and the state have always existed in Europe as distinct and opposing bodies. Although the state always maintains superiority in the hierarchy with the sovereign right to create others, the city has nevertheless existed as a mutually exclusive actor (Isin 2007, 215).

This differs from the traditional Chinese concept of politics (政, or more specifically, positive life (各正性命 *gezheng xingming*), which is itself a gift of heaven (天命 *tianming*)), emphasized incessant personal perfection in daily life. As everyone in fact lived in it, it was necessary to participate in disparate activities called politics. Accordingly, a container type of space for politics, such as *polis*, and concomitantly, an independent and self-regulated district separated from the state, was considered totally irrelevant. Instead a more diffused concept of politics favored a more dynamic setting, with every space as a becoming to a broader horizon (境域 *jingyu*). At the end, the totality of horizons, was the well-known concept of *tianxia* (天下, the Heaven). The concern of politics was not so much sovereignty as responsibility. Rather than concentrating numerous individuals, under various spatio-temporal circumstances, into a common container, the task was to mold each of them to become a responsible agent within his/her respective local setting (Chen 2007, 3–20). In short, traditional Chinese thought was not interested in rationally conceiving of spatialization, including the production of autonomous spatial containers.

This traditional thought received further development when the Qin dynasty underwent centralization by adopting the legalist style (法家). To centralize all decisions, the emperor implemented a spatialization in which every individual with his/her responsibility, but still without autonomy, would be tied together as a whole. Nevertheless, given the huge territory spanning under his throne, something unheard of by many sovereign states in Europe, the emperor had to rely on a system of local agents amidst the relentless centralization of decision-making, approving, legislating, surveilling and supreme judicial powers all under his auspices.

Following Cao and his associates (Cao 2011, 2018; Cao and Luo 2013), I call this a two-tier centralization-*cum*-delegation governing system (hereafter, two-tier system) (「上, 下分治」體制). Cao attributes the stability of the Chinese governing system over the past forty or so years to the establishment of a two-tier system. The center (including the provincial governments and their immediate subordinates) focuses on the hiring, promoting, supervising, assessing, and awarding and penalizing officials of the lowest rank ("the power of governing officials", the personnel-tier, P-tier), while the latter are empowered to administer people within their territories ("the power of governing people", the district-tier, D-tier). Having made this conceptual distinction, Cao (2011, 5–6; see also Cao 2018; Cao and Luo 2013, 1–2) comments that this two-tier system has in fact been established since the invention of the county system. Thus, basically since the Qin dynasty, there has been a two-tier system under the hegemony of the emperor. This hegemony was perpetuated by, first, centralizing all policy decision-making and assignment and allocation of officials at his disposal, and, second, relying on recruited local officials to govern on his behalf. Such a two-tier system – *xian*, on the one hand, and, on the other, its intermediate superiors – was like the skeleton interweaving the flesh, the territory, and was considered necessary and practical, given the empire's agrarian commercial economy.

This two-tier system exhibited three characteristics, rendering it distinguishable from administrative systems in the West. The first relates to a misunderstanding of centralization. It was a truism that *xian* was always on the receiving end of commands made elsewhere; it was never empowered to make momentous decisions while being locked into an extensive network of government. Nevertheless, this did not imply that a *xian* was left with no discretion (a term I prefer over autonomy) in governing its people. Since it was literally difficult to control everyday activities of individual magistrates within his territory, the emperor reluctantly ended up delegating the tasks of governing mundane local affairs to these officials. The latter in turn took advantage of the discretion and benefitted accordingly. While centralization with rigidity appeared to be the norm, delegation with discretion was the practice. This rendered the imperial system extremely complex – simultaneously practicing centralization and delegation – thereby rejecting any rigid, hierarchical representation of the division of administrative labor in simplistic concepts like center-local relations.

The second emphasizes the necessity to eschew a narrowly focused understanding of the formal structure. There were many more agents of substance other than those formally identified. Below *xian*, there were non-official self-organizations like *xiang* (鄉) (a township consisting of 10 *ting*), *ting* (亭) (10 *li* with a pavilion in-between), *li* (里) (100 households) and *shen* (什) (10 households) as well as territorial alliances. These self-organizations had given birth to the gentry. In addition, parallel to *sheng*, and above it too, were some offices like *Buzhengshisi* (布政使司, Minister of Civic Affairs), *Anchashisi* (按察使司, Minister of Financial Affairs), *Xunfu* (巡撫, Inspector-general) and *Zongdu* (總督, Governor), exercising power independent of those strictly empowered by the two-tier system. Again, these add-on features have blurred the boundaries of the spatial administrative hierarchy, thereby rendering the relationship between the administrative assemblage, the actual system, and its extension, something commonplace, extremely "symbiotic" (e.g., Wong 1999; Yi 2005, 114–128).

The third characteristic underscores the intertwining of natural and administrative forces. A *xian* consisted of two related activities: production and administration. One found within a *xian* numerous villages to accommodate the farmers, land and irrigation infrastructures for their grain production activities, and markets for the exchange of agricultural produces. These activities were mostly bound up with nature, its rhythms and forces. To administer these activities, it was also imperative to accommodate within a *xian* spaces of functionaries, including *baojia* (保甲communal self-defense system) for the surveilling of peasants, and residences not only for the magistrate, but also, sometimes, for intermediate superiors. These activities were all informed by the centralizing aspiration of the empire, or the man-made rhythms. It was the intertwining of these natural and man-made rhythms that formatted the two-tier system.

It is apparent from the above that within the jurisdiction of the emperor, there were many complementary opposites in operation, including centralization-delegation, assemblage-extension and natural-administrative forces. The dynamics of these opposites can be understood within the focus-field relationship with the emperor being the focus, extending the field to tie together many other agents, including *xian* magistrates. This focus-field relationship had always undergone continuous changes, depending on the exact intertwining of the aforementioned forces at any moment. While the emperor always explored ways to seek out the viable contexts to take control of the reins, *xian* supported the empire in their own ways. Although *xian* were part of the field with the emperor being the focus, they were foci in themselves, extending their own fields. This applied to other agents within the two-tier system too. There was thus focus within a field, and *vice versa*. Concomitantly, the overall balances of all the competing aspirations, from *xian* to their intermediate superiors, from the assemblages to the extensions, and from natural to administrative considerations, determined the fields as well as the foci. One example is the claim made by Yi (2005) that relentless, runaway corruption at various levels had finally caused the demise of the empire system in 1911.

That there was focus within a field, and, conversely, field within a focus, renders it difficult to employ currently available epistemology and methodology to understand the complexity. Cartier (2015) once urged us to adopt the recently popular conception of co-constitution to comprehend governing institutions. Golubchikov, Badyina, and Makhrova (2014), for example, have included socialist legacy and capitalism to understand the co-production of post socialist spatialities in Central Europe. Peck and Zhang (2013) did the same, proclaiming that China is variegated capitalism. Attempts of this kind have failed to fully recognize the prevalence of opposite and interpenetrating forces that breed the possibility for each to alter into the other. It is the continuity and change that count. While each of the two forces creates the other, each also lives within each other in the ongoing transformation.

It is also theoretically problematic to employ many current spatial concepts to characterize the focus-field relationship. The field, as the sphere of influence (Hall and Ames 1995, 273), is different from spatial reach (e.g., Allen 2003) in that fields and foci are never finally fixed. The space was not, as argued earlier, represented in Lefebvrian abstraction with one coding dominating in the play of resolving spatial contradictions, like the case of the City of London (Allen 2003, 160–66). Nor is it the case, as in the development of the London and the South East region, that divergent representations by various agents have contributed to the politics in and about space there (Allen and Cochrane 2010). What really counted in China, however, were not so much representations as the mundane duties that *xian* magistrates could comfortably perform within a concrete setting by taking into account local conditions. Besides, the power of reach of the emperor was neither intensive nor extensive in manner. In his treatise *Geographies of Power*, Allen (2003) proposes a topological understanding of state power against a more widely upheld topographical one. The topological understanding argues that the state spatiality exercises its hierarchical powers of reach outwards or downwards over space through mediated and real-time connections. There are different lines of authority, negotiation and engagement, criss-crossing one another in their distinctive rhythms and spatial practices. Political actors, public and private alike, as well as global, regional and local, are "lodged" within a region, rendering the state's hierarchical powers not so much rescaled or redistributed as reassembled – renegotiated, displaced and reimagined – in terms of spatial reach. In other words, while the concern is to bridge the gap erected by physical distance, proximity and reach are played in intensive manner. In contrast, a topographical approach holds that the greater the distance that powers are dispersed, the more spatially extensive is the reach of the state's authority. It is the extensive manner of territorial or networked capabilities that count (Allen and Cochrane 2010). The spatiality of DEB was different from the above characterization (Zhou 2005, 58–84). The crux of the matter is that in the two-tier system neither *xian* nor its intermediate superiors were autonomous units.

Xian as a space of dependent delegate in DEB

First, *xian* was a dependent delegate in the two-tier system. It was situated at the end of the line of authority without any possible substantive negotiation between the emperor, the focus, and itself, the field. On the one hand, the *xian* magistrate could not negotiate with the emperor on designated policies, debating, rejecting, ignoring, displacing or reformulating them; he merely implemented them accordingly, and, concomitantly, was evaluated on that ground. Nevertheless, being a focus in itself, *xian* would perform its duties in such a way to extend its own field. In other words, delegation to perform duties, rather than autonomy, actually ended up prevailing in these spaces. On the other hand, the emperor had to rely on *xian* to extend his spatial reach to every corner of the territory – to reach the peasants from afar, whose grain production as well as manpower contributed to the stability and prosperity of the empire. The larger the quantity of grain produced and of manpower conscripted, the more resources there would be at his disposal. This consideration should also be extended to the periphery, where *xian* were specifically established for defense purpose. The larger the number of *xian* being established and secured, the more spatially extensive the reach of the emperor would be. Thus, the development of *xian* spaces affected the development of the empire, yet *xian* were prohibited from being autonomous.

Second, *xian* was linked to dynamism elsewhere in the two-tier system. Intermediating *xian* and the emperor were superiors like *jun* (郡, prefectures), *dao* (道, circuits (after 997 it was called *lu* (路))) and *zhou* (州, prefectures). Unlike *xian*, these superiors were only concerned with the monitoring of the performance of *xian* magistrates for risk minimization: social risks (peasant rebellion), agency risks (runaway governing problems) and military risks (by invasion) (Cao 2018, 510–11). As peasant rebellion may span over many *xian*, the territory of the preexisting, intermediate superiors may have to be expanded to enhance effective maneuvering of sufficient combative resources. A case in point is that in the Han dynasty, an additional, provisional agent, *bu* (部), was established to overcome the feeble *jun* (Zhou 2005, 59–60). In developing the surveillance system, such as *dao* in Tang dynasty and *lu* in Song, they were set up in such a way to facilitate the transfer of assessment reports "upwards" to the emperor. Thus, the topography and accessibility were major determinants in configuring these surveilling districts in space (Zhou 2005, 66, 71).

Again, once additional intermediate superiors were set up for the above purposes, they, being foci in themselves, started to extend their own fields. This had been the case of the establishment of *Jiedu shi* (節度使, Military Commissioners) in *dao* and, later, *fangzhen* (方鎮, provincial military and administrative districts) in the Tang dynasty (Zhou 2005, 66–67). The emperor's reaction to this possible reduction in power was to reduce the chance of separatism while re-centralizing the authority power. In the Song dynasty, for instance, powers of Commissioners were substantially scraped by dispatching *Zhizhou* (知州, Manager of Prefecture) directly from the center, thereby displacing *Zhoucishi* (州刺史, Prefecture Prefects) as the main source of authority. Similarly, *Zhixian* (知縣, Manager of *Xian*) replaced *xian* magistrates at the level of *xian* (Zhou 2005, 70). In addition, the emperor assigned several Commissioners with different responsibilities and powers to the same area, sometimes with overlapping territories. As a result, no one, however powerful, was able to dominate any region for long.

It has been repeatedly emphasized that *xian* was the most stable administrative unit in the dynastic spatial hierarchy, while others, like *jun, dao, lu* and *zhou*, have been established and re-organized, or, even, abolished, over time. Stability aside, however, *xian* had undoubtedly enveloped within it more or less dynamism initiated at a distance. The dynamism – administrative restructuring of intermediate superiors – may be internally induced by some causes within *xian* in the first instance, followed by the emperor's external reaction to it. Or it may be caused by the interaction between the emperor and intermediate superiors. This stability-*cum*-dynamism was bundled together as a whole by the authority power for the cause of forming his own focus and field. It is then difficult to understand *xian* space in isolation, without situating it within the prevailing centralization-*cum*-delegation processes. Inside any *xian* space, there were these superior spaces, and *vice versa*.

Third, it is inappropriate to characterize *xian* and its superiors as belonging to lower or higher levels, as in the western rhetoric. Dating back to 12th or 13th century in the West, local governments had already obtained autonomy. Knowledge of and control by a group over the conditions of one's existence and survival are essential ingredients of a level (Elden 2004, 238). The absence of these conditions in imperial China prevented the social formation of a level. In DEB, the emperor had monopolized the authority power of administering personnel, protecting the territory and repressing any upheavals. Although

P-tier appeared "above" D-tier, they were not administratively one level above the other *per se*. The only difference between *xian* and its intermediate superiors was that while the former handled daily, mundane activities, the latter dealt with risk minimization. Their functions did not overlap, but were complementary to each other. Since they were established without autonomy, it was difficult to reduce them as distinct levels within the DEB hierarchy. *Xian* needed to be supplemented with its intermediate superiors, which, in turn, cannot be configured without the emperor. In other words, instead of emphasizing their distinctiveness as levels, we should underscore their interrelation.

Finally, the appearance, as well as disappearance, of some of the intermediate superiors could not be classified as rescaling. It is easy to comment, as the literature on rescaling would have us believe, that as there was devolution of decision-making powers to these intermediate superiors, state power would have been undermined. This was not the case for China, given the two-tier system. A vivid illustration can be found in the Song dynasty, when *lu* was set up above *zhou*. As soon as a *lu* was established, it was further divided into four parallel but mutually dependent departments: *Zhuanyun shi* (轉運使, responsible for finance), *Tidian xingyu shi* (提點刑獄使, judiciary), *Anfu shi* (安撫使, military affairs) and *Tiju changping shi* (提舉常平使, water conservancy and disaster relief). Independent of each other, they presided over various territories, while they were assigned to supervise *zhou* prefects and *xian* magistrates. Accordingly, one can easily conclude that the objective of this administrative arrangement was not so much delegation as centralization of power toward the emperor. The latter could control these superiors, while reducing their territorial or networked capabilities to renegotiate, displace and reimagine issues among themselves due to their territorial incompatibilities (Zhou 2005, 71). In short, the establishment of intermediate supervisors above *xian* cannot be conceptualized out of the context of DEB as some kind of rescaling or, as in Allen's reformulation, re-assemblage.

Cheng, under one jurisdiction, had never existed independent of *xian*

The above clarifications allow us to shed some light on the nature of *cheng* and its relations with *xian*. Instead of being a duality, *xian* and *cheng* were polarities with each being determined by itself and by every other. In particular, *cheng* and *xian* were mutually embedded in that they were part of each other and lived within each other.

Cheng was the walled component and space within *xian*. It was through the functionary offices (military, civil government, educational institutions) within *cheng* that the *xian* magistrate could form part and parcel of the centralization system as well as invoke its delegation power. It was there where he could negotiate with the gentry as well as surveil the farmers to ensure grain production and collect their taxes. Although the lion's share of activities in it were not actually involved in nature, they, being the chief market towns of their regions, were linked to it either for personal food consumption or for tax and tariff collection and management of granaries. The empire's authority could be reached, as well as kept out of reach by others, in *cheng*.

Being mutually embedded, *cheng* and *xian* always flew into each other in the ongoing transformation process. On the one hand, there were farming activities within *cheng* for its residents. The Chinese had attained the rights to change their place of residence for a long time. Since the Tang dynasty in particular, there were peasant families on the upward-mobility course or aspiring to it, living inside the wall and engaging in commercial and cultural activities. When these people got old or were released from official engagement, however, they usually retired in the countryside outside the wall (Mote 1977, 103, 117–18). On the other hand, non-farm activities inside *cheng* had relentlessly grown and, concomitantly, spilled over to the other side of the wall. The expansion of administrative offices, army barracks, temples, residences of officers and related functions of empire-building forced many other activities, also undergoing expansion, to take place outside the walls. Since these other activities were not closely associated with the administration, there was no such requirement of being bounded within the walls. Or, more specifically (as with Suzhou between the late Tang and early Southern Song periods), commercial activities preferred locating beyond the walls for more direct access to the major commercial trading routes, even while many areas within the walls remained derelict or agricultural (Mote 1977, 119; Xu 2000, 92).

Amidst these transformations prevailed continuity, the wall had continued to exist, first, materially, and, later, symbolically. Most of these walls were originally built to mark the presence of the rule of the empire. They existed as a physical sign of demarcating *cheng*, the protected area, from its surroundings, the unprotected area. The message was clear: life would go on under the rule of the empire, as the latter was as powerful and reliable as the walls. But as time proceeded, especially in times of turmoil, these walls were usually attacked, causing people to flee. Increasingly, they stood more for the symbolic meaning of the whole regime than for their actual physical function. Because of this, walls persisted over time (Xu 2000, 86–88). It was not until the Republican era that the pressure of spatial conflict over physical walls had reached such an uncompromising level that there were repeated calls to demolish some of them. It was this continuity and change that underlined the dynamics of *cheng*-cum-*xiang* rather than the more static conceptualization of, as proclaimed by urban historians like Mote (1977) and Xu (2000), the morphological features of an urban-rural continuum.

In light of the earlier argument that there were, inside *xian* space, intermediate superior spaces, a *cheng* could simultaneously contribute other roles to the perpetuation of the two-tier system. In his critical evaluation of scalar thought, Isin (2007, 212) has conceptualized the city in the West as the one comprising actual and virtual spaces. Actual space refers to permanent, solid and enduring things, making up the physicality and materiality of the city, while virtual, though actual too, refers to the association that exists beyond the actual bodies and things that constitute it. In imperial China, there were certainly actual spaces in *cheng* as it consisted of walls enclosing functionary, residential and religious activities. But, since *cheng* was never established to produce citizens, the virtual could never be *civitas* at all. Nevertheless, they were also virtual in the sense of Isin's classification as these solid things in *cheng* were related to the district magistrate of its contiguous *xian*, which was in turn intertwined with the emperor far away due to their inclusive relationship. These walls, as parts of a country-wide defense system, served the psychological function of marking the presence of the empire. Besides, there were within *cheng* other actual spaces associated with more than one administrative unit that existed far beyond the buildings and walls that constituted it. They were actual, as their activities constituted the physicality and materiality of the *cheng*, but under different subordination. They were also virtual in the sense that some of their activities were associated with administrative units located far away. Examples of this category included *fuguo xian* (附郭縣, appendage *xian*) – a *xian* without an independent administrative seat but existing as an appendage to that of an intermediate superior (Zhou 2005, 205). The administrative seats of Panyu *xian* and Nanhai *xian* are a case in point. They were once appendages to Guang *zhoufucheng* (see below). The difference between these two groups of *cheng* relate to their respective focus-field relationships in the two-tier system. For one, the field of influence was smaller, within the boundary of a contiguous *xian*. For the other, being related to the emperor and his associates, the field of influence extended well beyond the territory of a *xian*, spanning over *fu* or *zhou*. It is not uncommon for a *cheng* to be concurrently constituted by larger as well as smaller fields of influence.

In summary, *cheng* could neither stand on its own without the rest of *xian*, and *vice versa*. Nor could *cheng* possess defined powers of an independent territory, since it was part and parcel of *xian* for the functioning of the latter in the two-tier system. The relationship between *cheng* and *xian*, which depended on the intertwining of all associated forces over time, is all too complicated to be elaborated in simple dualistic terms or simplistic concepts like co-production or urban-rural continuum. The mutual embeddedness between *cheng* and *xian* should be more meticulously delineated as *cheng*-cum-*xiang* relations.

Cheng-cum-*xiang* relations: Guangzhou as an example

This discussion can be sharpened by focusing on the evolution of one city: Guangzhou *cheng* since its establishment more than 2,200 years ago. The city wall of Panyu *cheng* (番禺城), the embryonic form of Guangzhou *cheng*, was constructed in 214 B.C. by Ren Xiao, Governor of Nanhai *jun* (南海郡) (equivalent to today's prefecture), in the Qin dynasty. Since then, Guangzhou *cheng* has undergone a long and complicated development. In general, this local development corresponds to the evolution of the national two-tier system over time – the P-tier expanded or contracted so as to surveil the D-tier more effectively (Zhou 2005, 58–84).

The Qin dynasty employed the simplest two-tier system of *jun-xian*. Subordinated to the Nanhai *jun* was, among others, Panyu *xian*, which happened to also be the administrative seat. In 6 A.D., during the Han dynasty, *bu* (部equivalent to today's province) in the previously adopted *bu-jun-xian* system was replaced by *zhou* (州). Relevant to our discussion here was Jiao *zhou* (交州), which was later sub-divided into Jiao *zhou* and Guang *zhou* (廣州) in 226 (this was different from the contemporary Guangzhou in the administrative hierarchy). Guang *zhou* consisted of Nanhai *jun* and Panyu *xian*, among others, with the latter being the administrative seat for all three administrative units. During the Sui dynasty, Nanhai *jun* was abolished and reestablished as Nanhai *xian*, with the latter's territory exactly replacing Panyu *xian*. Nanhai *xian* thus appeared in the system for the first time. Being subordinated to Guang *zhou*, Nanhai *xian* also located its administrative seat in Guang *zhoufucheng* (附城; Guang *fu* was the abbreviation of Guang *zhou dudu* (廣州都督the Military-Governor) and *fu* (府the administrative seat of its own territory)). Again in 703, during the Tang dynasty, there was an administrative re-alignment to partition Nanhai *xian* and reestablish part of it as Panyu *xian*. During the Song dynasty, there were attempts both to incorporate Panyu into Nanhai and to separate the former from the latter. As both Nanhai and Panyu *xian* were under the auspices of Guangzhou *lu* (廣州路renamed from Guang *zhou* 廣州) in the system of *lu* (equivalent to province)-*fu/zhou-xian*, it was well documented that Guang *zhoucheng* (廣州城) had since been both governed by the two *xian*, more specifically with Nanhai taking over the reins of the western part and Panyu taking over the eastern part. In the Yuan dynasty, Guangzhou *lu* was under Guangdong *dao* (廣東道), which was in turn under Jiangxi *sheng* (江西省) in the complicated system of *sheng-dao-lu-xian*. Guangzhou *lu* had a territory covering 7 *xian* (including Nanhai and Panyu *xian*) and 1 *lushisi* (錄事司the register overseeing records of residents living in *xiang* (districts) within *cheng*). The Ming dynasty simplified the system to *sheng-fu-xian*, with Guangzhou *lu* renamed as Guangzhou *fu* (廣州府) (including one *zhou* and 15 *xian*) within Guangdong *sheng*. The Qing dynasty basically followed the Ming system, with Guangzhou *fu* ruling 14 *xian*. Guangzhou *cheng* remained governed by both Panyu and Nanhai until 1921, when Guangzhou was designated as an independent *shi* and, finally, 1932, when both Panyu and Nanhai had officially relocated their administrative seats to Xinzao (新造) and Foshan (佛山), respectively (Chen 1996, 44–50; GDCW (Guangzhoushi Difangzhi Bianzhan Weiyuanhui) Edited 1998, 139–147).

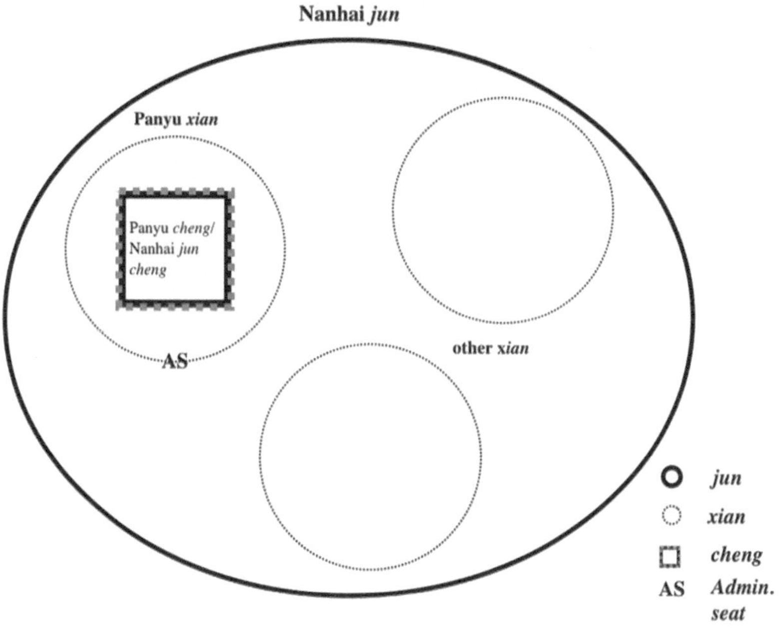

Figure 2. A Schematic Representation of Panyu *cheng* in 214 B.C.

One can draw a few implications from this *longue durée* of Guangzhou *cheng*. First, being the administrative seat of Nanhai *jun*, Panyu *cheng* was, in the first instance, central to the jurisdiction of Emperor Qin after he had conquered Lingnan (嶺南) (Figure 2). The mutually embedded *cheng*-cum-*xiang* relations have thus been in place for more than two thousand years. Second, the relations continued to develop over time to satisfy the changing needs of governing. As it was found necessary to consolidate the expanding territory, the growing economy and the exploding population, later dynasties, especially since the Song, complicated the spatial administrative hierarchy by installing additional intermediate superiors like *zhou, du, lu* and *fu*. Simultaneously, Nanhai *jun* had undergone expansion, contraction or even extinction, both forming Guang *zhou*, on the one hand, and, on the other, Nanhai *xian*, and annexing or abolishing Panyu *xian*. Third, the economy of Lingnan had grown enormously since the Ming, and Guangzhou *fu* started to incorporate even more *xian* than before. Figure 3 is a painting of part of Guangzhou *fu*, including some of its subordinating *xian cheng*, namely Nanhai and Panyu (inside Guangzhou *fu cheng* 廣州府城), and Dongguan (東莞), Zengcheng (增城), Huaxian (花縣), Qingyuan (清遠) and Sanshui (三水) (outside it). This painting also illustrates vividly that Guangzhou *fu cheng*, the actual, could not be discussed in isolation from its intertwining spatial relationships, the virtual, with the surrounding countryside. The relationship was more than that between an individual *cheng* and its immediate surrounding. Being the administrative seat of Guangdong *sheng*, Guangzhou *fu cheng* was interrelated with the countryside of other *fu*,

Figure 3. Guangzhou *fu* and Its Subordinating Units (painted between 1685 and 1729) (Source: modified from ZDLD (Zhongguo Diyi Lishi Dang'anguan, Guangzhoushi Dang'anyu and Guangzhoushi Yuexiuqu Renmin Zhengfu). Edited 2003, 23).

since all were under the auspices of the *sheng*. Within its own *fu* territory, Guangzhou *fu cheng* was immediately connected with the countryside of not only Nanhai and Panyu *xian*, which were partially located within its city wall, but also 12 others, which were elsewhere. While the *cheng*-cum-*xiang* relationships in the former two *xian* were grounded contiguously in physical space, those in the latter spanned over a wider and discontinuous space, sometimes via the relationship between the *fu cheng* itself and another *xian cheng*.

Fourth, the *cheng*-cum-*xiang* relationship between Guangzhou *cheng* and its immediately surrounding countryside had developed amidst the persistence of the city wall. According to Liu (1986), this phenomenon of having developments breaking away from the wall had taken place three times in Guangzhou *cheng*. Back in the Tang dynasty, overseas traders had already gathered in its western part. Then, beginning in the Southern Song Dynasty, development spilled over the wall and a commercial district flourished on the fluvial plain in the south. As soon as commercial activities developed to a comparable scale, a new wall extension was constructed to encircle it for governing purpose. Concomitantly, activities in the "Xincheng (新城 new *cheng*)" suffocated while Xiguan emerged as the international and local commercial and trade center during the late Ming and early Qing period, itself propelled by the closing of China to outside trade and, later, the declaration in 1759 of Guangzhou as the sole port for foreign

Figure 4. The Division of Subordination in Guangzhou, 1900 (Source: modified from ZDLD (Zhongguo Diyi Lishi Dang'anguan, Guangzhoushi Dang'anyu and Guangzhoushi Yuexiuqu Renmin Zhengfu). Edited 2003, 80)

trade within the whole country. By the beginning of the Qing, Xiguan had already become the prosperous center for merchant activities in Guangzhou *cheng* and, related to it, the handicraft industrial center. Figure 4 supports this argument in that developments in *cheng* had extended well beyond the city wall into other parts of Nanhai and Panyu *xian* themselves. Finally, Guangzhou *cheng* was clearly demarcated into two parts, with the western part being under the auspices of Nanhai *xian* and the eastern part belonging to Panyu *xian*. Even in the Tongzhi period of the Qing dynasty (1856–1875), there was marked division in the territory by streets, within the old city wall, within the new city wall and outside the city wall between the two *xian* (GDCW (Guangzhoushi Difangzhi Bianzhan Weiyuanhui) Edited 1998, 148–151). There was *cheng* in *xian* as well as, the reverse, *xian* in *cheng*. While centralization had designated Guangzhou *cheng* to govern the concerned territory on behalf of the emperor, as *cheng*-cum-*xiang hezhi* (integrated governing by *cheng* and *xian*), it was delegation, the mechanism by which its necessary *cheng*-cum-*xiang* morphological form and content were subtly constructed.

Summary and conclusion: toward a new understanding

To summarize, there was an intimate *cheng*-cum-*xiang* relationship during the dynastic past. This finding was developed by first reviewing the existing literature, which is found wanting. Most urban China research has drawn on western concepts of rural and urban, modern and tradition and city and suburb, and randomly indigenized them to depict town and country dualities in China. There are other attempts to remedy the situation by drawing on more local realities, but they have also failed because they either extrapolate current experiences backward to the past or simplify the underlying mechanisms to the single productive force.

This paper has adopted the methodology of spatial story, which is informed by *tongbian* philosophy. Theoretically, it has taken the dynastic past as the backdrop for a better understanding. To perpetuate the hegemony of one jurisdiction while promoting the agrarian economy, DEB had institutionalized a spatialization schema within a two-tier system. *Xian* was always on the receiving end of commands in the two-tier system, yet it had maintained sufficient discretion and benefitted accordingly. Besides the formal administrative structure, *xian* encountered many local self-organizations as well as superior offices independent of those empowered by the two-tier system. Finally, *xian* accommodated not only agricultural activities of natural rhythms but also activities informed by the centralizing aspiration of the empire. It was the interactions of these centralization-delegation, assemblage-extension and natural-administrative forces that substantiate the nature and activity of *xian*. It was a space of dependent delegation within the focus-field relations. *Cheng* was also actual and virtual in the workings of these relations. Accordingly, *cheng* could neither stand on its own without *xian*, and *vice versa*. Their relationship is too complicated to be elaborated in simple dualistic terms.

As noted earlier, it is beyond the scope of this paper to extend this discussion forward in history to the Republican era, and Maoist and post-reform China; attempts of this kind have to be left for future papers. In the recent literature, Cartier (2015) has coined the concept of territorial urbanization to denote the post-reform development (see also Cartier 2019). Wong (2019) in this issue in particular has helped make up this deficiency too by arguing somewhat along my thinking against dualism while focusing on the post-reform situation. Nevertheless, as it stands, the findings within this paper hope to enable us to ask interesting theoretical questions about, first, urban China research and, then, urban studies. For the former, it is common for the literature to randomly indigenize the western concept of the state, including state-society relations, to the Chinese situation. By venturing into the epistemological, methodological and theoretical wilderness, this paper proclaims that it is unnecessary as well as inappropriate to continue random indigenization. It has coined the concept of DEB to denote the distinguishable nature of the empire's perpetuation of the hegemony of one jurisdiction. This concept has also taken issue with many popular spatial concepts in the social sciences like spatial reach, level and scale. Instead, it emphasizes the concept of tier, which must be comprehended dynamically within DEB. Territory, too, must also be situated within the consistent perpetuation of one jurisdiction.

While the above theorization has shed light on the dynastic past, it may be easily extended to the socialist present. The methodology of spatial story, which narrates the development of China in space and time as a process of becoming, has equipped us with analytical tools as well as insights to decipher the present. The two-tier system of the spatialization schema of DEB is, according to Cao (2011), applicable to the socialist present too. What we need to do is meticulously conceptualize how the socialist regime has developed from the governing of "never-been-modern" during the Republican period to the current practices of orchestrating everything (see Figure 1), while extending the one jurisdiction even to the world by attempts like the implementation of the "One belt, One road" scheme and the promotion of 5G technology (Tang Forthcoming). By invoking the *tongbian* philosophy, I have avoided the dualism common in urban China research (see also Brown 2012). *Cheng* and *xian*, as polarities, are never dualistic as town and country. The former pair are mutually embedded in that *cheng* grows from *xian* and needs the other as partner, and *vice versa*. Duality identifies town and country as extrinsically related. In case their relationship is dissolved, they remain intact. In contrast, polarity refers to the relationship that is intrinsic and constitutive so that when it is cut, both parties will be affected seriously. Simply put, the effect in duality is *figuratively* affected while that in polarity is *literally* changed (Ames and Rosemount 1998, 24, emphases added). The subtleties enriched by the concept of mutual embeddedness are absent in other popular concepts like co-constitution or relationality. Besides, the mutual embeddedness between *cheng* and *xian* is, it must be reiterated, understood within the dynastic empire. This is achieved by situating them within the articulation of the focus-field relations of DEB. This is consistent with *tongbian* philosophy in that underneath the interaction of the polarities, or the parts, is the whole. This is something commonly ignored by the concept of duality.

The coinage of *cheng*-cum-*xiang* relations helps contribute to debates in urban studies, in which the town-country relation is one of the central debates. When Louis Wirth's "urbanism as a way of life" is combined with Robert Redfield's "folk society", the rural-urban continuum has occupied a prominent position in the discussion of the field. Intrinsic in town and country are their contradictions between each other. More recently, Henri Lefebvre adds fuels to the debate about dialectics as well as dualism by arguing that while there is complete urbanization under *mondialization*, the town-country contradiction has given way to the center-periphery. David Harvey has also elaborated town-country relations under globalization according to the logic of capital. It is fair to remark that town-country relations in urban studies have tended to concentrate on capitalist development under the western state. While Tang (2014b) has challenged such western bias by investigating the urbanization process in Hong Kong under colonialism, this study has continued the debate by contributing a different case study, epistemologically, methodologically and theoretically. While a full-fledged Chinese theory is still in the works, the concept of *cheng*-cum-*xiang* relations is the initial step or anchorage point to contributing to debates in urban studies. It is no longer justifiable to assert the existence of *the* concepts to understand urban in the world, nor the relevance of those crooked concepts like variegated or neoliberal urbanism under variegated capitalism.

Notes

1. See his own recent admission that the concept is inapplicable to China (Zhou, Lin, and Zhang 2018).
2. I have in mind a similar theorization of the Chinese state as Lebowitz (2012) has done for the Soviet Union. Preliminary attempts can be found in Tang's (1994; 1997, 32; 2006, 259–63; 2014a) concept of *étatization*. See below for more details on DEB.

Disclosure statement

No potential conflict of interest was reported by the author.

References

Allen, John. 2003. *Lost Geographies of Power*. Oxford: Blackwell.

Allen, John and Allan Cochrane. 2010. "Assemblages of State Power: Topological Shifts in the Organization of Government and Politics." *Antipode* 42: 1071–1089.

Ames, Roger T. and Henry Rosemount. 1998. *The Analects of Confucius: A Philosophical Translations*. New York: A Ballantine Books.

Brown, Jeremy 2012. *City versus Countryside in Mao's China: Negotiating the Divide*. Cambridge: Cambridge University Press.

Buckingham, Will. 2017. "Uncorking the Neoliberal Bottle: Neoliberal Critique and Urban Change in China." *Eurasian Geography and Economics* 58: 297–315.

Cao, Zhenghan. 2011. "Zhongguo Shangxia Fenzhide Zhili Tizhi Yiqi Wending Jizhi [China's Two-tier Administrative System and Its Stabilization mechanism]." *Shehuixue Yanjiu [Sociological Research]* 1, 1–40.

Cao, Zhenghan. 2018. "Centralization and Decentralization of Power Structure: A Theory of Ruling Risks and Empirical Evidence from Chinese History." *Chinese Journal of Sociology* 4: 506–564.

Cao, Zhenghan and Biliang Luo. 2013. "Jiquande Zhengzhi Fengjian Yu Zongjiang Fenquan – Cong Lishi Shijiao Kan Dangdai Zhongguo Shehui Guanli Tizhi Gaige [Political Risk of Centralization and the Vertical Separation of Power: Reform of Contemporary Social Management System from a Historical perspective]." *Nanfan Luntan [Southern Forum]* 2, 1–11.

Cartier, Carolyn. 2015. "Territorial Urbanization and the Party-State in China." *Territory, Politics, Government* 3: 294–320.

Cartier, Carolyn. 2019. "Exemplary Cities in China: The Capitalist Aesthetic and the Loss of Space." *Eurasian Geography and Economics* 60: 4.

Chan, Kam Wing. 1994. *Cities with Invisible Walls*. Hong Kong: Oxford University Press.

Chan, Kam Wing and Yanning Wei. 2019. "Two Systems in One Country: "The Origin, Functions, and Mechanisms of the Rural-Urban Dual System in China." *Eurasian Geography and Economics* 60 (4): 1–33.

Chen, Bin. 2007. *Tianxi Huo Tiandizhijian: Zhongguo Sixiangde Gudian Shiuyu* [Tianxia or between the Sky and the Earth: The Classical Horizon of Chinese thought]. Shanghai: Shanghai Shudian.

Chen, Daiguang. 1996. *Guangzhou Chengshi Fazhanshi* [The Development History of guangzhou]. Guangzhou: Jinan Daxue Chubanshe.

DeGlopper, Donald R. 1995. *Lukang: Commerce and Community in a Chinese City*. Albany: State University of New York.

Duara, Prasenjit. 1997. *Rescuing History from the Nation: Questioning Narratives of Modern China*. Chicago: University of Chicago Press.

Elden, Stuart. 2004. *Understanding Henri Lefebvre: Theory and the Possible*. London: Continuum.

Esherick, Joseph W. 2000. "Modernity and Nation in the Chinese City." In *Remaking the Chinese City*, edited by J. W. Esherick, 1–16. Honolulu: University of Hawaii Press.

Faure, David and Tao Tao Liu. 2002. "Introduction." In *Town and Country in China: Identity and Perception*, edited by D. Faure and T. T. Liu, 1–16. Basingstoke: Palgrave.

GDCW (Guangzhoushi Difangzhi Bianzhan Weiyuanhui) Edited. 1998. *Guangzhou Shizhi* [The Annals of Guangzhou shi]. Vol. 2. Guangzhou: Guanzhou Chubanshe.

Golubchikov, Oleg Anna Badyina, and Alla Makhrova. 2014. "The Hybrid Spatialities of Transition: Capitalism, Legacy and Uneven Urban Economic Restructuring." *Urban Studies* 51: 617–633.

Hall, David L. and Roger T. Ames. 1987. Thinking Through Confucius. Albany: State University of New York.

Hall, David L. and Roger T. Ames. 1995. *Anticipating China: Thinking through the Narratives of Chinese and Western Culture*. Albany: State University of New York Press.

Han, Guanghui, Yuejun Lin and Dan Wei. 2011. "Lun Zhongguo Gudai Chengshi Guanli Zhidude Yanbian He Jianzhi Chengshide Xingcheng [A Study on the Evolution of Urban Administrative System and the Formation of Designated Cities in Ancient china]." *Qinghua Daxue Xuebao* (Zhexue Shehui kexueban) [Qinghua University Bulletin (Philosophy and Social Sciences version)] 26: 58–65.

Huang, Philip C. C. 1993. "'Public Sphere'/'Civil Society' in China?: The Third Realm between State and Society." *Modern China* 19: 216–240.

Huff, Toby E. 2003. *The Rise of Early Modern Science: Islam, China, and the West*. Second Edition ed. Cambridge: Cambridge University Press.

Isin, Engin F. 2007. "City.Sate: Critique of Scalar Thought." Citizenship Studies 11: 211-228.

Kipnis, Andrew B. 2013. "Urbanisation in Between: Rural Traces in a Rapidly Growing and Industrialising County City." *China Perspective* 2: 5–11.

Lebowitz, Michael A. 2012. *The Contradictions of Real Socialism: The Conductor and the Conducted*. New York: Monthly Review.

Lefebvre, Henri. 1991. *The Production of Space*. Oxford: Blackwells.

Lewis, Mark Edward. 2006. *The Construction of Space in Early China*. Albany: SUNY Press.

Lin, Chun. 2006. *The Transformation of Chinese Socialism*. Durham: Duke University Press.

Lin, George C. S. 2007. "Chinese Urbanism in Question: State, Society, and the Reproduction of Urban Spaces." *Urban Geography* 28: 7–29.

Liu, Junde and Jinchao Fan. 2015. *Zhongguo Shizhide Lishi Yanbian Yu Dangdai Gaige [The Historical Evolution and Contemporary Reform of the City System in china]*. Nanjing: Dongnan Daxue Chubanshe.

Liu, Junde, Zuki Ma and Jing Xiong. 2012. *Zhongyang Zhixia Shizengqu Kongjian Zuzhi Yu Zhidu Moshi Tanxi – Lilun Jigou, Bijiao Fenxi Ji Shizheng Yanjiu*. [The Spatial *Organization and System* Model Analysis of Centrally-administered Shi Administrative District: Theoretical Framework, Comparative Analysis and Empirical study]. Nanjing: Dongnan Daxue Chubanshe.

Liu, Xin. 2002. "Urban Anthropology and the 'urban Question' in China." *Critique of Anthropology* 22: 109–132.

Liu, Zhiwai. 1986. "Mingqing Shiqi Guangzhoushe Jingjide Tese [The Characteristics of the Urban Economy in Guangzhou during the Ming-Qing period]." *Guangzhou Yanjiu [Guangzhou Research] No* 1: 62–65.

Ma, Shu-Yun. 1994. "The Chinese Discourse on Civil Society." *The China Quarterly* 137: 180–193.

McGee, Terry, George C.S. Lin, Mark Wang, Andrew Marton, and Jiaping Wu. 2007. *China's Urban Space: Development under Market Socialism*. Abingdon: Routledge.

Mote, F.W. 1973. "A Millennium of Chinese Urban History: Form, Time, and Space Concepts in Soochow." Rice University Studies 59,4. Accessed 10 January 2017. https:/hd1.handle.net/1911/63136.

Mote, F.W. 1977. "The Transformation of Nanking, 1350–1400." In *The City in Late Imperial China*, edited by G. William Skinner, 101–153. Stanford: Stanford University Press.

Peck, Jamie and Jun Zhang. 2013. "A Variety of Capitalism … with Chinese Characteristics?." *Journal of Economic Geography* 13: 357–396.

Solinger, Dorothy J. 1993. "China's Transients and the State: A Form of Civil Society?" *Politics & Society* 21: 91–123.

Sun, Shiwen. 2005. "Zhongguode Chengshihua Zhilu Zenmo Zou [Social Malfunctions of Urbanization in china)?]." *Chengshi Guihua Huikan* [Urban Planning Forum] 157: 9–17.

Tang, Beibei. 2015. "'not Rural but Not Urban': Community Governance in China's Urban Villages." *The China Quarterly*. doi:10.1017/S0305741015000843.

Tang, Wing-Shing. 1994. "Urban Land Development under Socialism: China between 1949 and 1977." *International Journal of Urban and Regional Research* 18: 392–415.

Tang, Wing-Shing. 1997. "Urbanisation in China: A Review of Its Causal Mechanisms and Spatial Relations." *Progress in Planning* 48: 1–65.

Tang, Wing-Shing. 2006. "Planning Beijing Strategically: 'One World, One Dream'?" *Town Planning Review* 77: 257–282.

Tang, Wing-Shing. 2014a. "Governing by the State: A Study of the Literature on Governing Chinese Mega-Cities." In *Branding Chinese Mega-Cities: Strategies, Practices and Challenges*, edited by P. O. Berg and E. Björner, 42–63. Cheltenham: Edward Elgar.

Tang, Wing-Shing. 2014b. "Where Lefebvre Meets the East: Urbanization in Hong Kong." In *Urban Revolution Now. Henri Lefebvre in Social Research and Architecture*, edited by L. Stanek, C. Schmid, and Á. Moravánszky, 71–91. Surrey: Ashgate.

Tang, Wing-Shing. Forthcoming. "Chinese Critical Geography: A Non-Dualistic, *Tongbian*-Informed Spatial Story." In *Placing Critical Geography*, edited by L. Berg, U. Best, and H. Larsen. Farnham: Ashgate.

Tian, Chenshan. 2005. *Chinese Dialectics: From Yijing to Marxism*. Lanham: Lexington Books.

Tsin, Michael. 1999. "Canton Remapped." In *Remaking the Chinese City*, edited by J. W. Esherick, 19–29. Honolulu: University of Hawaii Press.

Wong, Kit-Ping. 2019. "Behind the Chinese Urban Dualism: Territorially-Nested Urbanization – The Case of Dongguan." *Eurasian Geography and Economics* 60: 4.

Wong, R. Bin. 1999. "The Political Economy of Agrarian Empire and Its Modern Legacy." In *China and Historical Capitalism: Genealogies of Sinological Knowledge*, edited by T. Brook and G. Blue, 210–245. Cambridge: Cambridge University Press.

Xu, Yinong. 2000. *The Chinese City in Space and Time: The Development of Urban Form in Suzhou*. Honolulu: University of Hawai'i Press.

Yi, Zhongtian. 2005. *Diguo De Zhongjie: Zhongguo Zhuantong Zhengzhi Zhidu Pipan* [The End of the Empire: A Critique of China Traditional Political system]. Hong Kong: Sanliang Shudian.

ZDLD (Zhongguo Diyi Lishi Dang'anguan, Guangzhoushi Dang'anyu and Guangzhoushi Yuexiuqu Renmin Zhengfu). Edited. 2003. *Guangzhou Lishi Ditu Jingcui [The Essence of Guangzhou Historical atlas]*. Beijing: Zhonguo Dabaike Quanshu Chubanshe.

Zeuthen, Jesper and Michael Griffiths. 2011. "The End of Urban-rural Differentiation in China: Hukou and Resettlement in Chengdu's Urban-rural Integration." In *Politics and Markets in Rural China*, edited by B. Alpermann, 218–232. London: Routledge.

Zhou, Daming. 1997. "On Rural Urbanization in China." In *Farewell to Peasant China: Rural Urbanization and Social Change in the Late Twentieth Century*, edited by G. E. Guldin, 13–46. Armonk: M.E. Sharpe.

Zhou, Yu, George C. S. Lin and Jun Zhang. 2018. "Urban China through the Lens of Neoliberalism: Is a Conceptual Twist Enough?" *Urban Studies*. doi:10.1177/0042098018775367.

Zhou, Zhenghe. 2005. *Zhongguo Difang Xingzheng Zhidushi* [The History of China Local Administrative system]. Shanghai: Shanghai Renmin Chubanshe.

Territorially-nested urbanization in China - the case of Dongguan

Kit Ping Wong

ABSTRACT

This paper explores the question of extended urbanization by arguing for the need to overcome dualistic views to properly address the political specificities of urbanization in China. To this end, this paper understands state-territorial relations as a process of étatization by drawing on the particular literature on the role of the party-state in urbanization. Through a brief history of Guangdong, it elaborates on the political modalities of territorialization through China's administrative rank system. This has enabled the party-state to mediate the production of urban space. From this, I arrive at the concept of "territorially-nested urbanization", moving beyond limited accounts of hierarchical state powers in Chinese urban studies. Next, from a short periodization of Dongguan's urbanization, the paper exemplifies how a particular mode of territorialization has evolved into tense relationships between the city and towns in the ongoing dilemma of multi-centered versus concentrated direction of urbanization. Based on insights from in-depth fieldwork, the last part of this paper illustrates the contradictory mobilization of village collectives within extended state power through local government, and the development of villagers' politics and activism in contested land transformations.

Introduction

Dongguan, a prefecture-level city in the province of Guangdong, is an important case for understanding urbanization in China. The city has exhibited a different development form. In 2008, the city was badly hit by the global financial crisis, leading to numerous factory closures, capital flight and the exodus of massive numbers of laid-off migrant workers. To cope with this crisis, both the city and town governments have deployed various policies and planning strategies for economic and urban restructuring. In 2014, the State Council launched the scheme "Comprehensive Pilot Scheme for National New-Type Urbanization" (NDRC (National Development and Reform Commission) 2014). Humen and

This article has been republished with minor changes. These changes do not impact the academic content of the article.

Chang'an, the two wealthy industrialized towns of Dongguan, have been designated as two of 62 pilot areas to initiate this project. According to this scheme, Humen and Chang'an are to be promoted as "a new type of shi" (*xinxing chengzhen*), through "abolishing towns, establishing cities" (*chezhen sheshi*). Unlike the conventional way of designating a county-level city, the meaning of this "new type of city" was not specified in this document. Instead, it was generally speculated to be a "town-level city" (Interview, 2018). This administrative restructuring has granted these two governments a higher rank of power similar to that of the "county" level. After all, the scheme aims at re-engineering new rounds of urbanization, in which Humen and Chang'an have currently grown to a population of over 600,000, with GDP levels beyond the administrative power of a town government. By re-designating them as a new type of city, they would serve as two pioneers to propel further urbanization and economic restructuring in Dongguan.

The Pearl River Delta region, in which Dongguan is located, has been the subject of extensive scholarship characterized by popular approaches. These include the "bottom-up" or "village-led" and "state-led" or "top down" perspectives, in tandem with popular concepts such as *desakota*, as mentioned below. I would argue that these characterizations are problematic and misleading. In the next section, I will explain this and specify the political conditions of the Chinese urbanization. In particular, the question of the Chinese state is not defined by a dualist relation of top-down versus bottom-up, or center versus periphery, urban versus rural, nor can the case of Dongguan be understood as state-led versus village-led. Rather, to avoid thinking in dualist terms, this should be critically interrogated through the political modality of territorialization behind the process of urbanization. Ma (2005) has argued that the Chinese territory and urbanization are politically defined and organized by ranks (*dengji*) through the central control of the administrative division (*xingzheng quhua*) system. This specific process cannot be reduced to the general concept of scales. Furthermore, Cartier (2016), by coining the term "territorial urbanization" (Cartier 2015) and examining the fundamental conditions of ranks, has analyzed the roles of the party-state in designating cities, managing the production of all subnational territories, and reshuffling local governments' powers according to ranks, through which the state can coordinate the political economy of urbanization. Such analyses are crucially important to bring the intricacies into view. From these perspectives, the question of the Chinese state is thus defined not as a dualistic relation, but in relation to subnational territorial processes according to administrative ranks.

Based on the critical work of Cartier and Ma, I will elaborate on the political modality of territorialization in China in reference to the case of Guangdong province. Both scholars argue that the party-state can dominate both territorial urbanization and subnational governments at all ranks through the administrative division system. Specifically, it can accelerate as well as control new-city construction and urban restructuring for economic growth, while reshuffling all subnational governments in terms of their political ranks and powers. As such, in this paper, I will specifically explore an intertwined relationship between the production of space/territories and the state space/power. In China, the term territory is derived from the administrative division system. It refers to the institutional arrangement of all subnational territories and governments under the party's system of powers based on ranks. Yet, one needs to look beyond the institutional view, and therefore, here, I focus on a state-territorial relation not in terms of a thing (state-led), but as a process of étatization through the territorialization of urbanization and state power. This is the political modality in which Chinese urbanization is tied to a political ranking system that informs the production of different urban territories, mediates the hierarchization and differentiation of urban processes, and orchestrates all territorial dynamics related to concentration and extension of urbanization. The resulting reshuffling by the central state of local government powers and their territorial relations, which I coin here "territorially-nested political spaces", which in turn reconfigures and reproduces the state's space. Through time, these configurations have evolved and generated more complex local politics in the course of urbanization. Under this territorial nesting of political relations, different levels of the state are politically responsible to each other even as they challenge each other according to their own interests.

The next section explains how this intertwined relationship of urbanization and territorialization took place in Dongguan, and explores the process of "territorially-nested urbanization" among the city and towns, as well as subsequent contradictions of this process during further rounds of urbanization. The last section will examine these processes on the ground based on fieldwork conducted in Dongguan, especially in Tangxia town and surrounding villages, between 2014 and 2018. This fieldwork constitutes an important part of the study in revealing the urban as a mediation between diverse levels of force, processes and relations. It will become clear that state power does not operate in a clear, coherent or homogeneous way, but is, rather, opaque, multiple, and contradictory.

The chinese modality of (re-)territorialization: territorially-nested urbanization

To claim an urban concept for the Global South, the concept of *desakota* has appealed to many scholars to characterize the emergence of extended urbanized regions throughout Asia (Ginsburg, Koppel, and McGee 1991; McGee 2005, 2014). *Desakota* refers to villages (*desa*) and towns (*kota*), where in situ urbanization has taken place in many densely populated areas outside but close to cities or urban centers. This is the urbanization of the countryside that is characterized by an intensive mixture of agricultural and non-agricultural activities. Ginsburg (1991) and McGee (1991) recognize the *desakota* condition as the relative absence of distinction between rural and urban areas with tremendous changes in extensive rural areas outside cities in the postwar era. Yet its actual conditions have been over-generalized for Asian urban contexts. As Friedmann (2016, 163) argued, "Once researchers start to formulate more general theories of the periurban that go beyond place/time specificity, empirical evidence tends to be sparse." Also, by schematically classifying and demarcating space into three fixed geographical categories, namely, "peri-urban" and "*desakota*" surrounding a "city core" to form the model of an extended urban region, this geographical concept treats space as discrete and static forms and patterns, instead of critically theorizing and exploring these spaces as processes of urbanization. Thus, that approach would occlude us from interrogating specific conditions of, and differentiated processes of, extended urbanization in Asia. Lastly, this concept ignores political realities. As noted by Chan and Wei (2019, 15) in this issue, "The physical boundary between rural and urban areas become largely permeable, but the rural-urban divide based on the hukou has not." Chinese dualism is not explained by a quantifiable rural-urban boundary, but closely relates to the party-state regime, which can determine all essential means to (re)produce urbanization throughout the nation.

Moreover, "rural urbanization" or "periurbanization" in China tends to be equated with a bottom-up, or informal process of village-led urbanization, in contrast to a state-led one. A common way to analyze "the rural" is based on an analysis of discrete, dual-power relations, or a top-down and bottom-up relationship, between the state (or local governments) and villages. Outside the city proper, "rural" or "peri-urban" is not uncommonly imagined as an informal, unruly place waiting for a better governance to modernize and develop (Zhu 2013), as if villages were completely free from state control after 1978. Such an artificial boundary ignores a complex and historically intertwined relationship as rural collectivization had been able to fully extend state power into villages and peasant households (Buckingham and Chan 2018).

On the other hand, the notion of state-led has been problematic as well. The term "state-led" has come to signify "Chinese urban processes". Yet this notion is assumed as a given, fixed and homogeneous entity, and therefore is conceptually trapped in the criticism of state centralism (Brenner 1999). In practice, the term "state-led" can be ideologically constructed that assumes or mystifies state power, without questioning its continuity, rupture or contradictions.

Lastly, the concept of neoliberalism has been artificially applied to explain China's urban transformation (e.g. He and Wu 2009; Wu 2010). Again, in turn, it has been criticized as problematic and unjustifiable, for it decontextualizes China's specific political, economic and social structures (Nonini 2008; Tang 2014; Buckingham 2017; Zhou, Lin, and Zhang 2019). And thus Buckingham (2017) laments that there is neither clarification nor challenge to the basic assumptions of the state and markets behind the neoliberal claims, nor can these claims be justified to explain how China moved toward market operations when the state still has ultimate control over the political economy and society. In this regard, Zhou *et.al.* have pointed out that this concept is applied to show a "resemblance to the west" but leaves out the fundamental mismatches (Zhou, Lin, and Zhang 2019, 38). Also, in response, Tang (2014) and Cartier (2019) have raised the issue of the politics of knowledge production and have explained why the state space/power has not been taken seriously as the subject of the Chinese urban research. They challenge the random conceptual indigenization and appropriation through which many Western concepts are appropriated to explain government and planning in China, to the risk of emptying specific contexts of state histories and conditions of government (Tang 2014). This is also addressed by Cartier (2019) in this issue.

A critical analysis of the state should go beyond the questions of dual landownership and land development in China, although Lin (2009) has argued that land is important for understanding the complexities of China's political and social structures. In another vein, Ma (2005) reminds us that post-Mao China has undergone the rearticulation of, rather than the retreat of, state power in different forms of intervention at lower spatial scales. This leads me to consider state power *not* as a dualistic problem of land development as such, but as a dialectic relation between the state and its territory, through which specific political relations, urban processes and politics arise. An interrogation of the latter can unravel the political specificities of Chinese urbanization. In particular, Cartier's work (Cartier 2005, 2013, 2015, 2016), among others (Chan 2007, 2010; Fitzgerald 2002; Liu and Lin 2015; Ma and Cui 1987), are pertinent for my study to examine the state's territorialization in the course of extended urbanization. As mentioned above, Ma (2005) has argued that the notion of territory in China cannot be generally comprehended by the widely accepted concept of scales in urban studies but rather relates to the level of "sub-national territories" in accordance with the party's ranking system. Building upon Liu's (Liu and Lin 2015) work, Cartier has coined the term "territorial urbanization" (Cartier 2015), further problematizing specific territorial thoughts and practices present in the course of urbanization. Therefore, the drastic increase in the numbers and types of Chinese "cities" are not the common form of urban settlements many people assume they are, but a specific mode of state apparatus that should be understood as ongoing, large-scale, and systematic products of the state's territorialization.

The historical background to territorially-nested urbanization

Before moving to the case study, we should first distinguish the meaning of "cities" in China in order to understand how they have emerged in relation to the territorialization of state power (Cartier 2015). *Shi*, the translation of the English "city", has developed beyond a historic sense of walled cities (*cheng*) and a conventional sense of western cities. Guangzhou *shi* was promulgated in 1921 as an independent city council, separating itself from the historic rule of two counties.

In the 1920s, apart from three colonial territories (Hong Kong, Macau and Kouang-Tchéou-Wan), Guangdong had four designated cities: Guangzhou,

Shantou, Jiangmen and Foshan. Later, in the Mao era, *shi* were radically transformed according to the new party regime in favor of the centralization of state power. From 1949 to 1978, the number of designated cities increased to 11 in total. In terms of geographical territories, these cities were still based as the actual "urban settlements" and were institutionalized as "urban" to implement the national industrial strategy (Chan 1994). In terms of hierarchy, they were subject to a three-tiered administration at the provincial (*zhixiashi*), prefecture (*shangxiashi*) and county (*zhuanxiashi*) levels. In 1958, the central state adopted a new territorial strategy to put some counties under the leadership of larger cities in order to supply food to cities (Chan 1997, 91). As the provincial capital, Guangzhou started to extend its territory by establishing control over a larger countryside (*jiaoqu*) and incorporating counties from time to time. In contrast to cities, counties were "rural" and territorialized into a three-tier system of People's Communes to launch rural collectivization. It was through the total political control of these two separated economies, societies and registered populations that the state could continuously exploit peasants to supply cheap food and resources to cities and state-owned industries (Chan 1997; Chan and Wei 2019).

During the reform era, while the urbanization of national territory has continued to organize along the hierarchy of political rank, the state has adopted a different modality of territorialization to achieve new types of urbanization at different stages of reforms. This opened up a new era of production of large-scale new urban territories. *Shi* became fundamentally different from the previous "urban settlements" and were rendered multi-tiered "urban territories". Underlying these changes were the physical and political processes of territories. The first was the demarcation of territories and changes to administrative boundaries through merging, splitting, or annexing territories. The second were changes to government, diverting political power to city governments and eliminating rural territories, especially counties and prefectures (Chung and Lam 2004; Ma 2005), with the aim of integrating rural and urban areas into a unified mode of government and production. This modality of territorialization is not simply intended to enlarge material space for economic growth and accumulation, but also to sustain the party-state's étatization in the new era (Tang 1997, 2006, 2014). Through ranking, the central state can maintain a strong linkage between all levels of governments and particular levels of territorial urbanization for the sake of political domination.

Corresponding to different processes of territorialization, a political ranking system has been used to produce different scales and territories of urbanization in China. The national law designates different ranks of city-territories (*shi*) with particular territorial organizations (Chan 1997; Ma 2005). In reality, as shown below, territorialization of each city-territory should be understood as an ongoing and complex process of internal political restructuring. In the early 1980s, one territorial strategy was the production of third tier city-territories –

county-level cities – for the development of small cities. One approach was to give power to a county to develop into a city. In Guangdong, Zhongshan and Dongguan are examples of this process. Both were upgraded and territorialized almost completely from rural county areas. Another approach was to abolish and annex counties to expand county-level cities, e.g. Chaozhou and Meizhou. The purpose was to incorporate rural areas into the city's administration for urban development. These are legally known as "city-without-districts" (*bushiqu deshi*), which defines its territorial configuration to constitute lower-level administrative areas at the town level or below.

The state also adopted another strategy popular after 1983, "city-administering-counties", to exploit the rural sector. It abolished prefectures,[1] designated prefecture-level cities and put counties under cities' administration. These were all second-tier city-territories under the provincial government and involved the reterritorialization of larger jurisdictional areas, legally known as a "city-with-districts" (*shiqu de shi*). In 1983, Guangdong reterritorialized the pre-1978 prefecture-level cities by converting them from urban settlements into territorial ones. The eventual formation of six prefecture-level city-territories, together with Guangzhou, Shenzhen and Zhuhai, served as nodes of power to propel urbanization during the start of Guangdong's reforms.

Soon, the number of prefecture-level cities increased from 18 in 1988 to 21 in 1994. This round of changes aimed at eliminating all remaining prefectures and upgrading county-level cities and counties directly into prefecture-level cities. For example, the abolition of Meixian and Zhaoqing prefectures led to the upgrading of two county-level cities into prefecture-level ones. Huiyang prefecture was abolished and subdivided to form four prefecture-level cities: Huizhou *shi* and Dongguan *shi* were upgraded from county-level cities, while Heyuan *shi* and Shanwei *shi* were formed by merging several towns of counties. In 1994 Guangzhou and Shenzhen were designated as deputy-provincial cities, a half-rank higher than prefecture-level, but with economic planning directly under the central state. Thereafter, the number and boundaries of prefecture-level cities in Guangdong, made outside of provisions in the Constitution, have remained the same.

At lower ranks, however, between 1992 and 1996, a total of 34 counties were designated as county-level cities (Chen 2007) following the broadening of some designation criteria in 1993. All historic counties were abolished in Guangzhou, Foshan and Jiangmen. The purpose, in official words, was to accelerate the processes of rural-urban integration. In practice, this process was dubbed "city-making" (*zaocheng*) and corresponded to land speculation and development fever during the 1990s (Cartier 2001). Although these cities were constitutionally under the provincial leadership, they were politically (non-constitutionally) subjected to the administration of prefecture-level cities. Such a political arrangement is known as "city-administering-cities on behalf of the province" (*shidaiguan shi*). This form of political leadership, a higher-rank city administering a lower-rank city, soon led to further manipulation of administrative divisions by abolishing and converting many of these new cities into urban districts directly under prefecture-level cities (*cheshi shequ*). Through this strategy, Guangzhou, Foshan, and Shenzhen reshuffled and unified their internal territorial powers. The difference between being a county-level city or being turned into an urban district involves different processes of territorial urbanization. County-level cities have their own administrative, planning, and fiscal powers for the development of their territories. Promoting counties to county-level cities created new political centers to extend the overall urbanization process and, in effect, reduced the level of agglomeration of main cities. On the contrary, urban districts are directly governed by Guangzhou *shi* without separate administrative system and power. By turning them into urban districts, Guangzhou *shi* has been able to remove all government powers from these areas and integrated all of their land and population directly under its government and planning while new districts have been assigned specialized roles to realize a high-level strategy of the "Greater Guangzhou" in the city's Master Plan. In this sense, territorial powers have shifted and concentrated in prefecture-level cities.

The reshuffling of territories and governments cannot be normalized or treated merely as "institutional" arrangements. Instead, it demonstrates an intertwined relationship between the state's territorialization and urbanization as a historical and continuous process. As Cartier (2013) noted, this system raised the questions of state power and territory in Chinese urban theories. Also, the Chinese administrative division system reflects the spatial structure of the state, and it actually constitutes a crucial part of the party-state national development (Liu, Jin, and Zhou 1999). Thus, it diverges from Ma's (2005, 494) argument that this system "continues to be rigid and hierarchical" and "imposes a set of strong hierarchical power grids onto China's economic landscape". In other words, the territorial and urbanization processes on the ground are more complex. This is why an overemphasis on the rank relationship to analyze state power tends to underestimate its territorial reach and complexities in relation to China's urbanization process. Importantly, the territorialization of urban spaces and government relations leads to the formation of specific "territorially-nested political relations". However, my usage of "territorially-nested" is somewhat different from that of Neil Brenner (1998). While he conceptualizes a process of territorially-nested governance on the basis of capitalist territorial organization according to a set of general geographical scales (urban, regional, national, and global), I use this term to articulate a form of political organization and territorial relations based on the above-mentioned Chinese regime of territorialization, but not in capitalist terms. My usage of the term describes the territorial arrangement of state power in which *all* administrative areas are territorially-nested inside one another according to ascending ranks, and others are juxtaposed next to each other at the same rank. This way, the term usefully articulates the party's hierarchy of powers and responsibilities (*quanze*) in terms of both rank and territorial relationships between all levels of governments. The case of Guangdong illustrates that all rank-based city-territories are variously constituted by different nested forms of administrative areas.

Thinking this way, one can question how the state is able to exert its power. Below, I will address how different forms of political arrangements can lead to the formation of specific internal territorial powers and political relations by referring to the case of Dongguan. From an ethnographical perspective, the appearance of things and relations are usually different from what they actually are, seeming clear from a distance, but appear different in the field (Simone 2014). This is why an empirical field study is crucially important to explore what particular kinds of social relationships and specific territorial powers emerge in the process of urbanization. After all, the geography of power is neither absent nor homogeneous, but can be complex, multiple, and full of contradictions.

The multiple facets of territorially-nested urbanization: Dongguan

In the wider context of political territorialization in Guangdong, Dongguan became a prefecture-level city in 1988 as the result of territorial changes in two changing administrative divisions. In the case of Dongguan, a particular mode of territorialization has intertwined with specific processes of urbanization, instead of assuming it as the lack of state intervention, or labeling it as "bottom-up", "informal", "chaotic" village-led processes. This section first explores the specific way this city was established and politically territorialized. It is guided by questions of how the configuration of territorial powers co-evolved with particular processes of territorial urbanization and what kinds of territorial dynamics between the city and towns have been contradictory and constantly reworked.

Dongguan is unusual as a prefecture-level city directly administering towns and villages without districts or counties. As mentioned above, a prefecture-level city is "a city with districts" and administers counties or county-level cities. From 1988, the elimination of Huiyang prefecture gave rise to the formation of four prefecture-level cities, each of which had different political territorial configurations. Unlike the other three, Dongguan became a prefecture-level city without any counties for future territorial expansion. One could say that Dongguan county was "lucky" to become a prefecture-level city due to its location sandwiched between Hong Kong and Guangzhou, instead of being subsumed into Shenzhen, which has been hoping to acquire a larger territory. Moreover, Guangdong regarded Dongguan as an important locality for foreign investment via Hong Kong. In its total area of 2,465 km^2, Dongguan has only two levels of governments: a city and 28 towns. According to two former city leaders (Sanlianshenghuo 2008), such a territorial arrangement determined the internal political relations and powers in the course of Dongguan's urbanization. First, there is no county-level government to share profits or intervene in local powers; rather, economic power is given directly to lower-level governments. This arrangement greatly improves the efficiency and flexibility of the internal political decision-making process and also increases local productivity. Crucially, this is how the power relation between the city and towns is determined. If it would have retained a county-level government, the personnel appointment power would have remained with the province outside the city's scope. But since it has the power over the personnel appointment of the towns, it could politically recalibrate the city-town relations manifested in the subsequent geographies of multiple-power centers.

Specific to this governing structure, the 28 towns are granted certain powers equivalent to a county. These towns have been designated to Dongguan by the province since 1986 in order to develop "industrial satellites" parallel to its other cities. They are not old towns or market towns, which had no separated powers for development. In contrast, designated towns have their own party committee, government, and people's congress, with legal powers in budgeting, planning, and land development in their jurisdictions. Thus, without an intermediate county level, town governments have informally enjoyed a certain degree of administrative and economic powers, equivalent to those of the county, and all town party secretaries and cadres have taken on political and economic tasks to attract foreign investment, develop industries, and encourage urban construction.

The crucial point, however, is the mobilization of village collectives as a part of local government in Dongguan. There are 551 village collectives[2], which are legally defined as autonomous and independent of the government system. Yet, in reality, they are not autonomous and have become an extension of local government. In 1988, Dongguan adopted the governing system called "town-administering-villages". All these towns have become territorially based, transforming from rural districts (*chequ shezhen*, literally, "abolishing (rural) districts, establishing towns"), that were also former People's Communes. By establishing town governments, their territorial jurisdictions include town centers and a vast rural area of villages. Village committees were also village collectives, and were collectively organized along a two-tiered administration of brigades and productions teams. By integrating the administration of rural and urban areas under a town's leadership, these villages were designated as "management districts" (*guanli qu*), as a specific product of Guangdong policy (1987–1997). They served as an extension of local government to manage village land and affairs and played a key role in land development.

Given this specific mode of territorialization, Dongguan's urbanization has been tied to the development of the three-tiered political territories and relations. In the past of 30 years, this specific political configuration has enabled town governments and village collectives to develop together on their own terms and have attracted thousands of small industries and enterprises throughout its vast rural area. Dongguan's urbanization was characterized as "a spread of numerous stars in the sky without a moon" as described by local people (quoted in Lin 2006, 2011). Rural industrialization began from the 1980s, when towns and villages became localities of foreign investment from or through Hong Kong. Small workshops and factories started production in ancestral halls, communal canteens, town halls, and villages' offices, absorbing local surplus labor, and diversifying local household incomes. In the early 1990s, town and village leaders acquired large amounts of collectively owned farmland. Farmland was parcelized into industrial plots for lease through *sanlaiyibu*, along with new roads and basic infrastructures like water and electricity. Between 1990 and 1999, towns and villages constructed 255 industrial estates throughout Dongguan (Dongguan Daily 2014). In the late 1990s, the influx of Taiwanese industries greatly accelerated urbanization and the development of industrial clusters (Yang 2007). The migrant population increased from 1.45 million in 1997 to 5.87 million in 2006 (Dongguan Statistical Bureau 1998, 2007). This only shows the number of people recorded, which excludes a large number of "undocumented" migrants and their families, who worked in small workshops, hotels, food stalls, construction sites, recycling, etc. In 2010, five in 28 towns had more than 400,000 registered residents (Dongguan Statistical Bureau 2011).

In respect to the rationality of a rank-based political hierarchy, the relationship between the city and towns has not been so easy in the course of Dongguan's urbanization. This political configuration created internal contradictions and uneasy political relations between the two ranks of government during new rounds of territorial restructuring. First, due to its specific territorial governing structure, the spectacular expansion of industrialized towns has co-evolved with a high degree of multi-centered power structures of territorial development in Dongguan. On the one hand, these towns have played a crucial role in generating wealth and surplus and maintaining Dongguan at a high level of the GDP growth. On the other, for the towns' prepared master plans, it has been difficult for the city to coordinate their diverse interests or regulate their illegal constructions and redundant infrastructures. Second, the town-level administrative division has become a constraint on the city's further development. This problem is known as "a pony pulling a big cart", a situation where many towns have grown into large cities but governments have been limited by their town status to manage large population numbers, socio-economic activities, and the provision of required infrastructure, facilities, local welfare, and other services. Third, city leaders realized that "a city without centrality" was disadvantageous in promoting Dongguan for new rounds of PRD development. Thus, the urbanization of Dongguan has oscillated in tensions between strong multiple centers of towns as the engines of economic growth on one hand, and a stronger will of the city to seek new centralities and better coordination of the entire territorial development on the other. In contrast, the Shenzhen government has consolidated its territorial powers by transforming towns into districts, unifying its governing system in planning, land development, and management.

In response to the problems of territorial development, in 2001, Dongguan launched a project called the "Five-Year New City-Building Plan", as the first "card" to construct a new centrality (Lin 2006). Through some spectacular landscapes and architectural designs, the city government has built Nancheng, "the city of the South" as a new political, economic, and cultural center. The city has also expanded its center by developing Dongcheng, "the city of the East", through the construction of condominiums, shopping malls and hotels. Thereafter, the city began to construct the Songshan Lake Park and Guangdong Dongguan Eco-industrial Park. In the face of land shortage, additional land was acquired by transferring some areas from towns to the city and grouping them to develop hi-tech districts into new urban centers (Interview, 2018). Construction of highways, expressways, and railways has continuously reorganized the spatial relationships of Dongguan by strengthening all centers in relation to towns, developing the east-west corridor, and designating Humen and Changping as regional transportation hubs. Therefore, transportation has formed a grid network covering the entire territory, not only extending from south to north, but also from east to west.

Regarding the multi-centered structure of territorial powers, the city government proposed to change the administrative division of towns into districts. From 2006, the city began to suggest changes to town administration to resolve the problem of land scarcity. In 2010, the city launched a public discussion about a new role for Dongguan in the PRD's regional development, and presented the plan of "large urban districts", suggesting the merging of neighboring towns with the city center (Nanfang dushibo, August 16, 2011). But this plan has been contentious among the city and town governments and planners, for it was not simply about the spatial form of development, but the changes of powers and interests at stake in territorial development (Interview 2018). For the city, there were risks involved: towns have been the engine of the GDP and actual production sites, and thus careful consideration is needed in choosing possible towns to avoid negative impacts on the original sources of growth. For towns, it was about who should surrender their relative powers and interests of development to the city. Also, this plan was unfavorable to the towns in terms of the distribution of resources and interests as a large urban district would lead to a concentrated form of development. As such, this idea was likely to transform the future of Dongguan from a multi-centric territorial development to a center-peripheral one. The city could try to reshuffle its towns into "clusters" to balance out this problem, but the actual planning remains conflictual.

Territorially-nested urbanization on the ground: the case of town-administering-villages

As mentioned before, the third tier of territorial government comprises "town-administering-villages", in which village collectives have been absorbed into the local government in respect to urban construction and local administration. Hsing (2010) coined the term "village corporatism" by extending Oi's concept of "local state corporatism" to describe the roles of village collectives in distributional politics. Village collectives at "the rapidly developing urban fringe bargain with the urban government, profit from urban property markets, and define and defend their territorial autonomy under urban metropolitan governance" (*ibid.*, 6). She considers this a form of non-confrontational politics for economic rights to the land in the bottom-up process of "civic territoriality". She acknowledges different territorial logics and strategies of village collectives as a form of resistance and quest for citizenship rights in the face of land acquisition by higher level governments (*ibid.*, 83). However, this conceptualization of village collectives is not necessarily relevant to all places. In Dongguan, there have been different conditions arising from specific configurations of territorial powers, as argued above. It is also difficult to distinguish them as a form of "civic territoriality" that struggle for a certain degree of "territorial autonomy", since the political relations between the state/government and village collectives have historically been very complex. Also, village politics cannot be just taken as "village corporatism" or "collectives" as such, as it obscures the inherently complex, heterogeneous and conflictual relations in village collectives.

There is a strong sense of politics or political caution regarding land issues among local villagers as these issues reconfigure existing patterns of accessibility and land possession. Therefore, villagers could suspect me of being a spy from the media, government, or a foreign political unit. The collection of information from villagers also shows inconsistencies and divergences in their stories, partly reflecting how they learnt or acted through rumors or gossip in a context of political circumstances without accountability and transparency. Some villagers were also unable to speak openly because of a fear of possible retaliations.

Extension of the state power

In Tangxia, the town government and village collectives have risen to be two-tier territorial powers in the course of Dongguan's urbanization and industrialization process. Although the People's Communes were abolished in 1982, the village collectives are still organized by the second and third tiers of collective ownership: brigades (*dadui*) and production teams (*shengchan dui*). Village collectives are a political space in which rural areas are still tied to state power through village organization and collective landownership. Following the establishment of Dongguan *shi*, the political arrangement of "town-administering-villages" was established in which village collectives were designated as "management districts" and functioned as a part of local government power in adherence to Guangdong's policy. This policy empowered the positions of village party secretaries, who were positioned directly under town leadership, and management districts in charge of attracting of foreign investment, land management and construction, and population management and public security. Given this new political arrangement, there was a reshuffling of power within village collectives in the name of unified village leadership. Collective land was previously managed by production teams in charge of leasing farmland and allotting housing land to village households. From the late-1980s to the mid-1990s, village leaders retrieved a large amount of leased farmland from village households while transferring land management to the brigades in almost all village collectives. As a result, land allocation decisions were entirely controlled by village leaders. In 1997, the management districts were supplanted by village committees and village teams, and they were again changed into residential committees and small teams in 2004. Villages also introduced the share-holding system, election of village directors, and set up supervision teams. As shown below, the dominance of village party secretaries has not changed much.

Today, there are a total of 19 village collectives managing 152 village teams and respective village households in Tangxia. State power is evident throughout the entire political landscape. There are red banners everywhere praising the "socialist new village construction" set forth by Xi Jinping. Villages' boundaries are clearly demarcated, and many village collectives moved into new administrative buildings that replicate the design of party and government buildings at higher levels. Enterprises, migrants, and even villagers have long regarded them as local governments, and these administrative buildings are the conspicuous symbols of state power. At the time of my fieldwork, the village collectives had become more institutionalized as local government offices with budgetary powers over things like construction and development, building safety, migrants' residence cards, resolving labor disputes, road numbering, policing, and social harmony in their jurisdictions. Many of the village collectives have erected "gated" communities (*shequ*), with exclusive uses over facilities and infrastructures. These high-rise village housing estates, known as "*nongmin* (peasant) apartments" are now symbols ("face construction", *mianzi gongcheng*, as mentioned by a villager) of the "new socialist village construction" that are modernizing the villages.

Powers of circumvention and materialization

Nevertheless, being incorporated as the extension of the state power does not mean that village collectives fully commit to achieving the goals set forth by the central government or act in accordance with national laws. In this territorially-nested political space, it has been important for local governments or units align themselves to the state in terms of a unified party ideology. Yet, behind the veil of political ideology, there has been always struggle and manipulation to materialize different interests and spatial configurations. Given the enormous revenues gained from land development, all village leaders adopted different strategies and practices to turn the farmland under their jurisdictions into built-up areas. From 1988 to the mid-1990, Tangxia underwent a peak period of land acquisition to use farmland for industrial use and housing. According to villagers, some farmland[3] was lost to the town government for urban development, but most were lost to their village leaders in the name of collective interests, with only small amounts of compensation.

In the last thirty years, the urbanization of village territories has transcended the dual landownership system with land neither state-owned nor collectively-owned. In the 1990s, Tangxia's village development began to follow certain common practices. First, large areas of agricultural land were acquired from village households. After forming a large area of land, housing plots were demarcated along both sides of roads or alleys, while the rest of the interior land was divided into different industrial plots and internal roads. Nowadays, these residential and industrial areas form a neighborhood unit: multi-storey residential buildings and shops are located on either side of streets, surrounding many factories in the center. Each village collective has an average of two to four industrial estates. Before, these industrial and residential lands were sold by village collectives plot by plot. Many industrial lands were sold to foreign and domestic investors in the form of a 50-year land lease[4] which can be transferrable to other actors without changes in land use. It is a very simple lease contract. Lessees pay a cash fee within two years and can construct factories and dormitories while the village collectives take on the responsibility for land formation and infrastructure provision. Most housing land was divided into different sizes of plots and leased to villagers and non-villagers at a fixed price, without specific formulation of lease conditions. Some properties were later transferred to third parties that run cheap hotels or workers' dormitories, for example. And thus, the lease price varies for each village and changes over time. The sales of housing plots were administered through various means, for example, by holding auctions for the highest price, conducting balloting for plots, or requesting deposits (10,000 yuan) to participate in auctions. In a few cases, the plots were redistributed for compensation for farmland land loss, or after a conflict between village leaders and villagers.

Owners constructed multi-story houses on their plots, usually four to five stories, with shops or small workshops on the ground floor. Later, village collectives started to construct multi-story houses themselves for the sale of "minor property rights" (*xiao chanquan fang*). It was also common to first lease land and complete construction and apply for official permits and certifications at a later stage. The village collective would issue a document to a buyer who could build a house. Once a building was constructed, a land certificate could be obtained from the (town-level) Land Management Office. At that time, village collectives also provided a one-stop service to buyers to apply for land certificates at the town government. Later, however, these documents became invalid and villagers could also not get land certificates, although they could get building permits. Then, the application for both certificates and permits of housing were suspended, but villagers continued to build houses under the premise that "this is fine if it is only two-storied construction." But even the construction of larger housing blocks, up to 5-to-8 stories, were commonly seen in the field. While some owners of buildings hold both certifications of land-use and property right, many houses still have no certification, and many industries are only registered under the name of village collective land.

These processes, however, cannot be simply categorized as "illegal" or "black market," as suggested by Lin (2009). According to China's constitution, it is certainly illegal to develop collective land without a change to state-landownership. If they were regarded as illegal at that time, how could most of the rural land in Tangxia and even Dongguan be completely transformed without a high-level intervention? As noted by Yang and Wang (2007; 2008), Dongguan did not implement the national land regulations, such as land construction quotas or the control of farmland conversion. In fact, only after 2000 did the city government and planners start to carry out regulations to distinguish and control illegal constructions at the town and village level. Prior to this, urbanization took place in a *gray zone*, neither legal nor illegal. Urbanization proceeded in circumvention of power, where policies were initiated by the city and town governments that turned out to be legitimate changes on the ground. However, policies and regulations were constantly changed. Many were only implemented for a limited period of time, changed into another project, or suddenly terminated.

Tactics of multiplication and maximization

The massive scale of land development and investment on the basis of collective landownership did not necessarily take the form of village corporatism as conceived by Hsing (2010). The latter assumes the homogeneity and collectiveness of a village collective. In the case of Tangxia, all village collectives have hitherto been internally hierarchical, heterogeneous, interdependent, and, also, conflictual. The development of village territories was not really based on collective consensus among village households but on power relations among the town, village leaders, and villagers. While this political arrangement of town-administering-villages has reproduced the powers of domination and circumvention found in these village collectives, it has also given rise to a particular village politics in the course of urbanization.

Village collectives are internally hierarchized in the three-tier organization of village committees, small teams and village households. The small teams' administrative power in the early-1990s was vested in the village party secretary who stands at the head of village committees (that is the former brigades). They are the first leader of a village and they are not selected by their village fellows but by the members of local party committee. In other words, they are "chosen" by the town leader who determines the income and annual bonus of village leaders based on their yearly duties and achievements, known as the "village cadre assessment". According to Zhao's (2019) research in Shandong, this relationship between local officials and village cadres can be understood as "invisible rewards". The former allows the latter to profit from various forms of rents (e.g. extraction of profits from land acquisition or construction projects) within their jurisdictions, which can explain why the position of village party secretaries in Tangxia attracted entrepreneurial villagers, and why they could sell or lease land during the 1990s the way they did. These invisible rewards are in exchange for the villager leaders' compliance to the town leadership, including complete fulfillment of all assigned political and economic duties. This relationship between two levels of leaders, together with local developers, eventually developed into a complex network of interests in local land development. And yet, the relationship also depends on the internal dynamic of a village. For instance, a village collective is composed of many village teams, which could act as a counter power to the decisions of the village leaders. In response, the village collective has adopted different land allocation methods to compensate the village teams for their lost farmland on which they could build houses and factories.

As mentioned above, there was a strong sense of political awareness by the villagers. Given the political circumstances and power relations between town and village leaders, it is crucial to explore the politics of local villagers and their actions in the course of urbanization. Many villagers were illiterate but not apolitical or simply victims of power. Rather, they have developed a sense of consciousness and activism in the emerging land politics in Tangxia. For instance, a respected village cadre who had been in charge of land reforms in 1953 led his villagers to the town government in the mid-1990s. They objected and complained about the village party secretary selling a large number of housing plots and demanded a fair redistribution of land for villagers to build their houses. Eventually, the town leader finally made a concession in order to pacify villagers. As a result, some of the housing land was redistributed to all households at a "village price", and others were sold at a "market price". Until this day, many villagers feel thankful to this old village cadre, but the cost was great as he lost his position in the party committee. This incident indicates that the town leader had allowed the village leader to sell housing land (and industrial land) in the village. Many villages' leaders were selling housing plots at unaffordable prices at the time. There were many unresolved conflicts over land and complex power relations in villages as some villagers could not easily mobilize themselves as a "collective" force. Even in the case of secret land deals, petitioning by villagers seemed a promising tactic at the city level, but quickly turned to series of threats and retaliations at the town and village level.

Nevertheless, village households were learning to cope with these decisions and turn them to their own advantage. And, although Article 62 of the Land Management Law stipulates the rule of "one household, one homestead plot", numerous households have secured multiple landholdings for themselves over the years. Selling land on a plot-by-plot basis, choosing buyers by ballot, in situ land bidding, or conducting sales at night time provided specific platforms for certain groups of people to purchase "permanent" land leases. Local villagers paid attention to these practices in order to not miss out on land sales and development. Relying on locally-obtained information, local villagers would immediately take out their savings, borrow money, or call relatives in Hong Kong to bring back cash when they expected plots to be put up for sale. In this sense, this required them to pay attention to everything happening around and know the villagers involved in the process of land information and sales. Only then they could multiply housing plots and therefore obtain land rights in the village.

Another tactic to secure land rights is to maximize the number of crops and buildings to increase compensation when land is appropriated. Most villagers in Tangxia could only afford to build houses after they received compensation from the town government, so they deliberately maximized compensation by planting lychee trees on a hill where the town government was rushing through its plan to build a new town center in the early 2000s. There was no collective mobilization and organization, but rumors of the plan encouraged many villagers to participate in this tactic. The strategy was successful and villagers received a large compensation, around one million yuan, from the town government which they used to construct houses and buy cars. As a villager said, this opportunity was "a turning point of changing livelihoods" in that village. Yet, when the same tactic was followed in another village, the government changed the rules to limit the amount of compensation.

Building houses is also a basis for securing land rights. Initially, building a house was thought of as an improvement of village livelihoods, or used by their sons after marriage, or as a later source of income through leasing. Many villagers learned to build more houses in old village areas, for example, turning a courtyard or a former area for drying crops into rental houses. These houses look similar to old village houses, both of which are poor in terms of quality and lack sanitation. Villagers rent rooms cheaply to migrants for additional income. Most importantly, they have these houses in order to secure their land plots. They have been reluctant to give these housing plots back to village collectives, even though this is required to buy a unit of the *nongming* apartments. Building more houses is a popular strategy for strengthening villagers' negotiating position in the process of land acquisition, especially in new rounds of urban renewal initiated by either the town government or village committees. During an urban renewal project in 2003, town officials attempted to stop villagers from building houses for more compensation, but villagers, alongside each other, continued to construct houses on their land, trying to maximize their opportunities to negotiate with the town government. This lasted for two years and led to the breakdown of negotiations and the end of the project. Even though the city government reformed the village organization and carried out the shareholding system, binding all villagers into an identity of shareholders, local villagers have learned different ways of doing politics in an ongoing process of spatial production and the reproduction of domination in Tangxia.

Concluding remarks

Synthesizing and building upon the critical work of Cartier and other scholars, I have elaborated a political modality of territorialization, especially exploring a kind of political relationship in which the formation of specific territorial powers is linked to particular processes of urbanization. To illustrate this political modality, I have examined three different levels of the territorialization process: Guangdong, Dongguan city and towns, and Tangxia town and villages. I have argued that existing concepts, such as *desakota*, state-led or village-led urbanization, and top-down or bottom-up frameworks cannot explain the specificities of extended urbanization in China. Instead, I suggest that the analysis of state-territorial relations is a necessary starting point for understanding a specific intertwined relationship between territorialization and urbanization. In this analysis, the political modality of territorialization unfolds a complex relationship between the state, territory, and urbanization. First, through changes to administrative divisions, China's urbanization is tied to a political rank system that results in the production of city-territories, with their multi-tiered, rank-based and specific territorial configurations. Second, this process of territorialization is a political tool of the central state in producing different territories of urbanization and in reshuffling all sub-national governments and their territorial power relations. Third, I used the term "territorially-nested" to describe and explore particular spatial forms of state power arising from specific modes of territorialization as well as the process of urbanization on the ground. Taking this state-territorial perspective can reveal the changing geographies of state power, the formation of different territories and power relations, and their dynamics and contradictions in relation to urbanization.

Chinese cities were made to be administrative divisions, and as the state apparatus, Guangdong described here shows the production of different rank-based city-territories, the formation of internal territorial powers, and the ongoing processes of restructuration at different times. In the 1980s and 1990s, the number of county-level and prefecture-level cities increased dramatically, and they rapidly extended, diversified and differentiated more strategic areas for regional urbanization. This territorial strategy after 2000 has clearly shifted to the concentration of territorial powers to prefecture-level cities, forming a larger quasi-metropolitan area by subsuming county-level cities into urban districts. Through rank, changing administrative divisions of these cities have continuously redefined their political status and promoted their specific levels and pathways of urbanization, for example, from a county to a county-level city, and to a prefecture-level city or an urban district; or even directly from a county to a prefecture-level city. These city-territories are ranked and also hierarchized into a specific realm of "territorially-nested" power relations. Given that each of them are internally nested by other administrative divisions, changes in one of these areas will lead to changes in the internal power dynamics within and across these city-territories.

Instead of a rigid and hierarchical form of state power, this process of territorialization is full of complexities and dynamics of state power, and its relation to the dynamics of urbanization over time. The case of Dongguan showed the uneasy and contentious relationship between the city and towns arising in the processes of territorialization and urbanization. Given its specific form of territorial governing system, the prefecture-level city without districts, towns and villages have become the main driving force of "rural" industrialization. In the past 30 years, these two-tiered territorial powers have co-evolved with decentered, extended processes of urbanization. Dongguan faced problems in its next step of development in both concentrated and decentered processes of urbanization. The city developed different strategies to extend its power over the entire territory, but how to coordinate many of the super-large towns remained a problem. While these towns have reached their limit of development due to their administrative divisions, the city could not subsume them to form a large city district as a representative of new Dongguan in the PRD. Thus, this problem of government remained uncertain and unsettled in the relationship between the city and towns in Dongguan.

The fieldwork in Tangxia town and villages further explored the processes of territorially-nested urbanization and their contradictions on the ground. The case showed that the role of village collectives has functioned as an extension of local state power. This is quite different from that of Shenzhen, in which the administrative role of village collectives was eventually supplanted by street governments through several rounds of territorialization and étatization. In Tangxia, from management districts to the current residential committees, village collectives have institutionalized and expanded the administrative structure to govern their territories. Given the territorial arrangement of "town-administering-villages", village collectives were incorporated as a part of local government, but simultaneously created specific dynamics and politics of urbanization on the ground. This is not only due to the power of circumvention and manipulation of town and village leaders emerged in the specific political relationship, but also the emergence of political consciousness among villagers in the struggle of land acquisition and development. This political sense led to different actions of individual villagers, for example, multiplying their landholding, maximizing their compensation, or negotiation in the process of urbanization. This may not be a collective form of resistance. Rather many villagers have acted together side by side and circulated information through rumors or gossip that enabled them to participate and reshape the politics of urbanization and made significant impacts on the production of material space in their villages.

Notes

1. These old prefectures were the intermediate agents of the province to manage counties and were not set up as a local government.
2. The number of village collectives refers to the "management districts" (rural). Some of them were converted to residential committees since 2004, but the basic functions of collective ownership have remained intact.
3. I take the term farmland to refer to land used for different agricultural activities, e.g. paddy and vegetable fields, orchards, fish and poultry farm.
4. The lease term can vary according to agreements. A common form of land lease is 50 years, but there are also 60-year or 20-year leases.

Acknowledgments

This research is funded by a research project Territories of Extended Urbanisation carried out in the framework of the ETH Future Cities Laboratory Singapore. Earlier versions of this paper which provided an ongoing progress of this work, were presented at the Holcim Forum 2016 and the AAG 2017 Conference in Boston. I express special thanks to the project leaders Milica Topalovic and Christian Schmid, and the project team members Elisa Bertuzzo, Rodrigo Castriota, Nancy Couling, Alice Hertzog-Fraser, Hans Hortig, Nikos Katsikis, Metaxia Markaki, Philippe Rekacewicz, AbdouMaliq Simone, who contributed substantially to the development of the concept discussed here. Furthermore, I would like to especially thank for insightful and critical comments from Kam Wing Chan, Carolyn Cartier, Wing-Shing Tang and Geerhardt Kornatowski on the last version of the paper. All errors and omissions remain the responsibility of the author.

Disclosure statement

No potential conflict of interest was reported by the author.

Funding

This research was supported by the Singapore National Research Foundation and ETH Zürich.

References

Brenner, Neil. 1998. "Between Fixity and Motion: Accumulation, Territorial Organization, and the Historical Geography of Spatial Scales." *Environment and Planning D: Society and Space* 16 (4): 459–481. doi:10.1068/d160459.

Brenner, Neil. 1999. "Beyond State-Centrism? Space, Territoriality, and Geographical Scale in Globalization Studies." *Theory and Society* 28 (1): 39–78. doi:10.1023/A:1006996806674.

Buckingham, Will. 2017. "Uncorking the Neoliberal Bottle: Neoliberal Critique and Urban Change in China." *Eurasian Geography and Economics* 58 (3): 297–315. doi:10.1080/15387216.2017.1350196.

Buckingham, Will, and Kam Wing Chan. 2018. "One City, Two Systems: *Chengzhongcun* in China's Urban System." *Journal of Contemporary China* 27 (112): 584–595. doi:10.1080/10670564.2018.1433576.

Cartier, Carolyn. 2001. "'Zone Fever', the Arable Land Debate, and Real Estate Speculation: China's Evolving Land Use Regime and Its Geographical Contradictions." *Journal of Contemporary China* 10 (28): 445–469. doi:10.1080/10670560120067135.

Cartier, Carolyn. 2005. "City-Space: Scale Relations and China's Spatial Administrative Hierarchy." In *Restructuring the Chinese City: Changing Society, Economy and Space*, edited by L. J. C. Ma, and F. Wu, 19–33, London: Routledge.

Cartier, Carolyn. 2013. "What's Territorial about China? From Geopolitical Narratives to the 'Administrative Area Economy'." *Eurasian Geography and Economics* 54 (1): 57–77. doi:10.1080/15387216.2013.791198.

Cartier, Carolyn. 2015. "Territorial Urbanization and the Party-State in China." *Territory, Politics, Governance* 3 (3): 294–320. doi:10.1080/21622671.2015.1005125.

Cartier, Carolyn. 2016. "A Political Economy of Rank: The Territorial Administrative Hierarchy and Leadership Mobility in Urban China." *Journal of Contemporary China* 25 (100): 529–546. doi:10.1080/10670564.2015.1132771.

Cartier, Carolyn. 2019. "Exemplary Cities in China: The Capitalist Aesthetic and the Loss of Space." *Eurasian Geography and Economics* 60 (4).

Chan, Kam Wing. 1994. *Cities with Invisible Walls: Reinterpreting Urbanization in Post-1949 China*. Hong Kong: Oxford University Press.

Chan, Kam Wing. 1997. "Urbanization and Urban Infrastructure Services in the PRC." In *Financing Local Government in the People's Republic of China*, edited by C. P. W. Wong, 83–125. Hong Kong; New York: Oxford University Press.

Chan, Kam Wing. 2007. "Misconceptions and Complexities in the Study of China's Cities: Definitions, Statistics, and Implications." *Eurasian Geography and Economics* 48 (4): 383–412. doi:10.2747/1538-7216.48.4.383.

Chan, Kam Wing. 2010. "Fundamentals of China's Urbanization and Policy." *China Review* 10 (2): 63–93.

Chan, Kam Wing, and Yanning Wei. 2019. "Two Systems in One Country: The Origin, Functions, and Mechanisms of the Rural-Urban Dual System in China." *Eurasian Geography and Economics* 60 (4).

Chen, Chao. 2007. *Zhongguo Xingzheng Quhua Yange Shouce* [Handbook of the Evolution of the Chinese Administrative Divisions]. Beijing: China Map Publishing House.

Chung, Jae Ho, and Tao-chiu Lam. 2004. "China's "City System" in Flux: Explaining Post-Mao Administrative Changes." *The China Quarterly* 180: 945–964. doi:10.1017/S0305741004000694.

Dongguan Statistical Bureau. 1998. *Dongguan Statistical Yearbook*. Beijing, China: China Statistical Press.

Dongguan Statistical Bureau. 2007. *Dongguan Statistical Yearbook*. Beijing, China: China Statistical Press.

Dongguan Statistical Bureau. 2011. *Dongguan Statistical Yearbook*. Beijing, China: China Statistical Press.

Dongguan Daily. 2014. "Dongguan Xian Gongyou Gongye Qu 503 Ge [Dongguan Has 503 Industrial Estates]." 27 February. Accessed 10 January 2016. http://epaper.timedg.com/html/2014-02/27/content_1242191.htm?div=-1

Fitzgerald, John. 2002. *Rethinking China's Provinces*. London; New York: Routledge.

Friedmann, John. 2016. "The Future of Periurban Research." *Cities* 100 (53): 163–165. doi:10.1016/j.cities.2016.01.009.

Ginsburg, Norton. 1991. "Extended Metropolitan Regions in Asia: A New Spatial Paradigm." In *The Extended Metropolis: Settlement Transition in Asia*, edited by N. Ginsburg, B. Koppel, and T. G. McGee, 27–46. http://www.jstor.org/stable/2059045?origin=crossref

Ginsburg, Norton, Bruce Koppel, and Terry. G. McGee. 1991. *The Extended Metropolis: Settlement Transition in Asia*. Honolulu: University of Hawaii Press.

He, Shenjing, and Fulong Wu. 2009. "China's Emerging Neoliberal Urbanism: Perspectives from Urban Redevelopment." *Antipode* 41 (2): 282–304. doi:10.1111/j.1467-8330.2009.00673.x.

Hsing, You-tien. 2010. *The Great Urban Transformation: Politics of Land and Property in China*. Oxford: Oxford University Press.

Lin, George C. S. 2009. Developing China: Land, Politics and Social Conditions. Routledge Contemporary China Series. London: Routledge.

Lin, George C. S. 2006. "Peri-Urbanism in Globalizing China: A Study of New Urbanism in Dongguan." *Eurasian Geography and Economics* 47 (1): 28–53. doi:10.2747/1538-7216.47.1.28.

Lin, George C. S. 2011. "Urbanization of the Pearl River Delta: The Case of Dongguan." In *China's Urban Space: Development under Market Socialism*. 1st ed., edited by T. McGee, G. C. S. Lin, M. Wang, A. Marton, and J. Wu, 96–120. London: Routledge.

Liu, Junde, Runcheng Jin, and Keyu Zhou. 1999. *Zhongguo Zheng Qu Dili [China's Administrative Geography]*. Beijing: Sciences Press.

Liu, Junde, and Tuo Lin. 2015. *Zhongguo Xingzhengqu Jingji Yu Xingzheng Quhua [The Chinese Administrative Economy and Administrative Division: Theories and Practices]*. 1st ed. Nanjiang: Southeast University Press.

Ma, Laurence J. C. 2005. "Urban Administrative Restructuring, Changing Scale Relations and Local Economic Development in China." *Political Geography* 24 (4): 477–497. doi:10.1016/j.polgeo.2004.10.005.

Ma, Laurence J. C., and Gonghao Cui. 1987. "Administrative Changes and Urban Population in China." *Annals of the Association of American Geographers* 77 (3): 373–395. doi:10.1111/j.1467-8306.1987.tb00165.x.

McGee, Terry. 2005. "Distinctive Urbanization in the Peri-Urban Regions of East and Southeast Asia Renewing the Debate." *Jurnal Perencanaan Wilayah Dan Kota* 16 (1): 39–55.

McGee, Terry. 2014. "The Emergence of *Desakota* Regions in Asia: Expanding a Hypothesis." In *Implosions/Explosions: Towards a Study of Planetary Urbanization*, edited by N. J. Brenner, 121–137. Berlin: Jovis.

McGee, Terry. 1991. "The Emergence of *Desakota* Regions in Asia: Expanding a Hypothesis." In *The Extended Metropolis: Settlement Transition in Asia*, edited by N. Ginsburg, B. Koppel, and T. G. McGee, 3–26. Honolulu: University of Hawaii Press.

Nanfang dushibao. 2011. "Dongguan Chengshi Zongti Guihua Jiang Xiu Bian, Zhongxin Chengqu Kuorong Re Zhengyi [The Controversies of the City Expansion in Dongguan]." 16 August. Accessed 5 June 2016. http://www.xxghj.gov.cn/news/Detail.aspx?sid=3c286e91e2158fd9.

NDRC (National Development and Reform Commission). 2014. "The Notice on the Comprehensive Pilot Scheme of the National New Type of City/Town." 29 December. Accessed 10 May 2018. http://www.ndrc.gov.cn/gzdt/201502/t20150204_663081.html

Nonini, Donald M. 2008. "Is China Becoming Neoliberal?." *Critique of Anthropology* 28 (2): 145–176. doi:10.1177/0308275X08091364.

Sanlianshenghuo. December 11 2008. "Cong Xiangcun Dao Chengshi [From Villages to a City]." Accessed 10 July 2016. http://news.sina.com.cn/c/2008-12-11/102316827446.shtml

Simone, AbdouMaliq. 2014. *Jakarta: Drawing the City Near*. Minneapolis; London: University of Minnesota Press.

Tang, Wing-shing. 1997. "Urbanisation in China: A Review of Its Causal Mechanisms and Spatial Relations." *Progress in Planning* 48: 1–65.

Tang, Wing-shing. 2006. "Planning Beijing Strategically: 'One World, One Dream'?" *Town Planning Review* 77: 257–282.

Tang, Wing-shing. 2014. "Governing by the State: A Study of the Literature on Governing Chinese Mega-Cities." In *Branding Chinese Mega-Cities: Policies, Practices and Positioning*, edited by P. O. Berg and E. Bjorner, 42–63. Cheltenham: Edward Elgar Pub.

Wu, Fulong. 2010. "How Neoliberal Is China's Reform? the Origins of Change during Transition." *Eurasian Geography and Economics* 51 (5): 619–631. doi:10.2747/1539-7216.51.5.619.

Yang, Chun. 2007. "Divergent Hybrid Capitalisms in China: Hong Kong and Taiwanese Electronics Clusters in Dongguan." *Economic Geography* 83 (4): 395–420. doi:10.1111/j.1944-8287.2007.tb00380.x.

Yang, Daniel You-Ren, and Hung-Kai Wang. 2007. "The Regional Variations of Farmland Property Rights Transformation in China: An Institutional Comparison between Suzhou and Dongguan." *Issues & Studies* 43 (4): 117–148.

Yang, You-Ren, and Hung-Kai Wang. 2008. "Land Property Rights Regimes in China - A Comparative Study of Suzhou and Dongguan." In *China's Emerging Cities: The Making of New Urbanism*. 1st ed., edited by F. Wu, 26–43. London: Routledge.

Zhao, Tan. 5 October 2019. "The "Invisible Reward." Paper presented at the 2019 AACS Annual Conference, University of Washington, Seattle, USA.

Zhou, Yu, George C. S. Lin, and Jun Zhang. 2019. "Urban China through the Lens of Neoliberalism: Is a Conceptual Twist Enough?" *Urban Studies* 56 (1): 33–43. doi:10.1177/0042098018775367.

Zhu, Jieming. 2013. "Governance over Land Development during Rapid Urbanization under Institutional Uncertainty, with Reference to Periurbanization in Guangzhou Metropolitan Region, China." *Environment and Planning C: Government and Policy* 31 (2): 257–275. doi:10.1068/c11309.

Index